Natural Language Dialog Systems and Intelligent Assistants

Gary Geunbae Lee • Hong Kook Kim
Minwoo Jeong • Ji-Hwan Kim
Editors

Natural Language Dialog Systems and Intelligent Assistants

Springer

Editors
Gary Geunbae Lee
Dept. of Computer Science and Engineering
Pohang University of Science and Tech
Namgu, Pohang, Korea,
Republic of (South Korea)

Hong Kook Kim
School of Information and Communications
Gwangju Institute of Science and Tech
Buk-gu, Gwangju, Korea
Republic of (South Korea)

Minwoo Jeong
Microsoft Corporation
Redmond, Washington, USA

Ji-Hwan Kim
Sogang University
Department of Computer Science
and Engineering
Mapo-gu, Seoul, Korea
Republic of (South Korea)

ISBN 978-3-319-19290-1 ISBN 978-3-319-19291-8 (eBook)
DOI 10.1007/978-3-319-19291-8

Library of Congress Control Number: 2015948371

Springer Cham Heidelberg New York Dordrecht London

Printed on acid-free paper

Springer International Publishing AG Switzerland is part of Springer Science+Business Media (www.springer.com)

Preface

The International Workshop on Spoken Dialog Systems (IWSDS) 2015 was held in Busan, Republic of Korea, from January 11 to 13, 2015.

This workshop series brings together researchers from all over the world working in the field of spoken dialogue systems. It provides an international forum for the presentation of research and applications and for lively discussions among researchers as well as industrialists. On the success of IWSDS'09 (Irsee, Germany), IWSDS'10 (Gotemba Kogen Resort, Japan), IWSDS'11 (Granada, Spain), IWSDS'12 (Paris, France), and IWSDS'14 (Napa, USA), this year's workshop has designated *Dialog System and Intelligent Assistant* as a special theme of discussion. We encouraged discussions of common issues of spoken dialogue systems including but not limited to:

Speech recognition and synthesis
Speaker/language recognition
Spoken language understanding
Dialog management
User modeling/simulation
Evaluation of dialog system
Multi-modality/emotion recognition from speech
Question answering from speech
Speech data mining
Language resource and databases
Machine learning for spoken dialog systems
Educational and healthcare applications

The workshop program consisted of 23 regular papers and 7 demo papers. In particular, we enjoyed two keynote talks of *Dr. Mazin Gilbert*, Assistant Vice President, Intelligent Services Research, AT&T Labs, USA, and *Dr. Haizhou Li*, Director of Research, Institute for Infocomm Research, Singapore. In addition, we had one tutorial by *Dr. Gary Geunbae Lee* from POSTECH, Korea, one panel discussion, and four sponsor talks. This book gathers revised versions of selected papers presented at the workshop. They cover the topics such as:

Personal assistant with dialog systems
Proactive and anticipatory computing
Dialog systems connected to knowledge base
Big data for large scale spoken dialog system

We would like to take this opportunity to thank the IWSDS Steering Committee and the members of the IWSDS'15 Scientific Committee for their timely and efficient contributions and for completing the review process on time. In addition, we would like to express our sincere gratitude to the members of the IWSDS'15 Local Committee who contributed to the success of this workshop with careful consideration and timely and accurate action.

Pohang, Republic of South Korea — Gary Geunbae Lee
Gwangju, Republic of South Korea — Hong Kook Kim
Redmond, WA, USA — Minwoo Jeong
Seoul, Republic of South Korea — Ji-Hwan Kim
March 2015

Contents

Chapter 1
Rapidly Scaling Dialog Systems with Interactive Learning

Jason D. Williams, Nobal B. Niraula, Pradeep Dasigi, Aparna Lakshmiratan, Carlos Garcia Jurado Suarez, Mouni Reddy, and Geoff Zweig

Abstract In personal assistant dialog systems, *intent models* are classifiers that identify the intent of a user utterance, such as to add a meeting to a calendar or get the director of a stated movie. Rapidly adding intents is one of the main bottlenecks to *scaling*—adding functionality to—personal assistants. In this paper we show how *interactive learning* can be applied to the creation of statistical intent models. Interactive learning (Simard, ICE: enabling non-experts to build models interactively for large-scale lopsided problems, 2014) combines model definition, labeling, model building, active learning, model evaluation, and feature engineering in a way that allows a domain expert—who need not be a machine learning expert—to build classifiers. We apply interactive learning to build a handful of intent models in three different domains. In controlled lab experiments, we show that intent detectors can be built using interactive learning and then improved in a novel end-to-end visualization tool. We then applied this method to a publicly deployed personal assistant—Microsoft Cortana—where a non-machine learning expert built an intent model in just over 2 h, yielding excellent performance in the commercial service.

Keywords Language understanding • Natural language processing • Spoken dialog systems • Spoken language understanding • Machine learning • Machine teaching • Interactive learning • Active learning

---

Work of the authors "Nobal B. Niraula" and "Pradeep Dasigi" was done while at Microsoft Research.
The authors "Jason D. Williams", "Nobal B. Niraula", and "Pradeep Dasigi" contributed equally to this work.

J.D. Williams (✉) • A. Lakshmiratan • C.G.J. Suarez • G. Zweig • M. Reddy
Microsoft Research, Redmond, WA, USA
e-mail: jason.williams@microsoft.com

N.B. Niraula
University of Memphis, Memphis, TN, USA
e-mail: nbnraula@memphis.edu

P. Dasigi
Carnegie Mellon University, Pittsburgh, PA, USA
e-mail: pdasigi@cs.cmu.edu

G. Geunbae Lee et al. (eds.), *Natural Language Dialog Systems and Intelligent Assistants*, DOI 10.1007/978-3-319-19291-8_1

1.1 Introduction

Personal assistant dialog systems are increasingly a part of daily life, with examples including Microsoft Cortana, Apple's Siri, Google Now, and Nuance Dragon Go. Figure 1.1 shows a high-level pipeline for these systems. First, spoken input is recognized using an open-domain automatic speech recognizer (ASR) and converted to words, such as "Am I free today at noon?." This step is skipped if input is provided by text. Next, the input words are processed by intent detectors to infer the user's intent, such as READFROMCALENDAR. In parallel, entity extraction identifies utterance substrings that contain entities, such as the date "today" or "noon," and entity resolution maps those substrings to canonical forms, such as `2014-09-11` or `12:00:00Z-08:00:00`. Finally, a function is called that takes the intent and entities as input, optionally updates an internal state, and produces a response as output. The cycle then repeats. Although variations exist—for example, intent detection and entity extraction/resolution may be done jointly—these are the high-level components of personal assistant dialog systems.

To add new functionality to state-of-the-art commercial systems, only certain stages of the pipeline need to be modified. On the one hand, the ASR service generally remains unchanged, because the coverage of modern ASR platforms is quite broad. Also, it is often the case that the function calls to produce a response are already available—for example, on most mobile phone platforms, APIs already exist for manipulating a user's calendar information. On the other hand, adding a new intent almost always requires building a new intent detector model. New entity extractors are only sometimes needed, because many entity types can be reused for new intents: for example, the intents to *read*, *write*, and *delete* an appointment from a calendar all share the same entities: times, dates, locations, and so on.

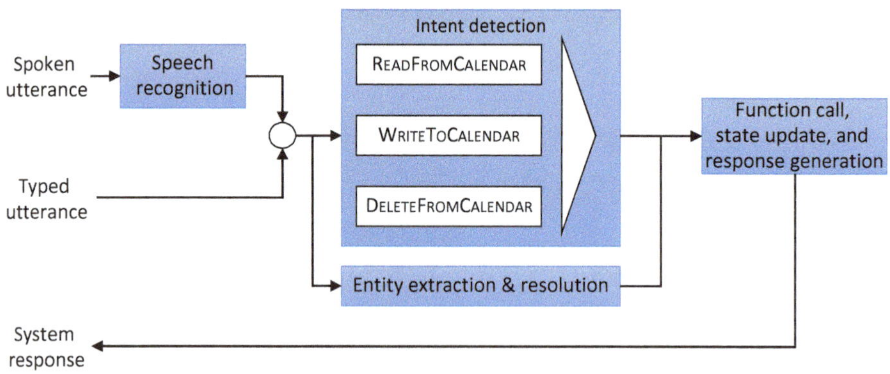

Fig. 1.1 Processing pipeline used for personal assistant dialog systems. First, utterances which are spoken are converted to text. Intent detection and entity extraction/resolution are then performed on the text. The resulting intent and entities are used to select and call a function. This function optionally updates an internal dialog state, then produces a response. The cycle can then repeat

In sum, quickly building new intent detector models is the key step in adding new functionality to a personal assistant dialog system. Yet building new intent detectors is often a slow process. One reason is that typically many people are involved, requiring coordination and scheduling. For example, a *data engineer* collects utterance data that contains instances of the target intent; a *user experience designer* creates a labeling instruction document that explains the new intent; a *crowd-source engineer* creates a crowd-sourcing task where workers apply the labeling instructions to data; and a *machine-learning expert* uses the data to build an intent detection model. Another problem is that issues with the definition of the intent often surface only at the end of the process when model performance is measured, requiring the whole loop to be repeated. Overall, the entire process can take weeks.

In this paper we introduce a new method for building intent detector models that reduces the time required to a few hours. The work is done by a *single person*, who is an expert in the domain of interest, but is *not* an expert in machine learning. The key idea is to use *interactive learning* (Simard et al. 2014), which interleaves intent definition, active learning, model building, model evaluation, and feature engineering (described in detail in Sect. 1.3).

This paper is organized as follows. The next section reviews intent detection and related work, then Sect. 1.3 introduces interactive learning and explains how it is applied to building a single intent model. Section 1.4 presents an evaluation, then Sect. 1.5 introduces and evaluates an end-to-end tool that enables developers to correct any stage of the intent/entity processing pipeline. Section 1.6 describes a live deployment in Microsoft Cortana of an intent model built using interactive learning. Section 1.7 briefly concludes.

1.2 Background and Related Work

Intent detector models are classifiers that map from a sequence of words to one of a set of predefined intents—for example, from "Am I free this afternoon" to READFROMCALENDAR, which is one of the predefined intents (Wang et al. 2005; Tur and Mori 2011). A typical domain like calendaring has on the order of a dozen intents. In this paper, a *binary* classifier is trained for each intent, which allows new intents to be built and tested independently, facilitating extensibility across domains. Intent i is a model of the form $P_i(y|\mathbf{x})$, where $\mathbf{x}$ are the words in the utterance and y is a binary variable where $y = 1$ indicates that the intent is present in the utterance and $y = 0$ indicates not. For a given utterance $\mathbf{x}$ and a set of intents $\mathscr{I}$, the most likely intent i^* can be selected as

$$i^* = \arg\max_{i \in \mathscr{I}} P_i(y = 1|\mathbf{x}) \tag{1.1}$$

Out of domain utterances $i = \varnothing$—i.e., those which match none of the intent detectors—can be explicitly modeled with a background model $P_{\varnothing}(y = 1|\mathbf{x})$.[1]

The model itself can be estimated in a variety of ways, such as boosting (Schapire and Singer 2000), support vector machines (Haffner et al. 2003), or deep neural networks (Sarikaya et al. 2011) among others—and the focus of much past work has been to maximize performance (accuracy, F-measure, etc.) given a fixed dataset. The approach described in this paper admits any model class which can be trained rapidly, but for simplicity we have used regularized log-linear models. Features will be words, n-grams of words, or other simple lexical features, such as the length of the utterance.

The approach in this paper focuses on maximizing performance for a given *time budget*, where time can be spent on labeling or feature engineering. The primary time-budget approach taken in past work has been *active learning* (Tur et al. 2003, 2005). Active learning starts with a small seed set of labeled data instances, from which an initial classifier is trained. This classifier then scores a large set of unlabeled instances. A selection rule then draws instances based on their scores. For example, the rule might draw examples for which the classifier shows greatest uncertainty, e.g., a score of 0.5. As more instances are labeled, the classifier is retrained, and the process is repeated. As compared to random sampling, active learning has been shown to have much better time efficiency, i.e., active learning requires fewer labels than random sampling to attain the same level of performance. By contrast, in this paper, we apply *interactive learning*.

Enabling nonexperts to quickly build data-driven dialog systems is a long-standing goal in the research literature (Glass and Weinstein 2001; Jung et al. 2008; Fukubayashi et al. 2008). Unlike past efforts, this work draws on a massive set of utterances from a deployed commercial personal assistant and builds upon interactive learning, described next.

1.3 Interactive Learning

Interactive learning (IL) is a method for efficiently building classification models, where classifier definition, labeling, model building, and evaluation are all interleaved and done by a single developer (Simard et al. 2014). Like active learning, IL is suitable when unlabeled data is abundant but labeling is expensive, as in our case. IL incorporates active learning but extends it substantially.

IL requires a large database of unlabeled data instances, such as webpages, emails, or (in our case) text of utterances to a personal assistant. The database

[1]This approach assumes that the scores are directly comparable. In this paper, the classifiers are not guaranteed to produce comparable scores, but since only a handful of classifiers are used and their calibration is similar enough, this mismatch will not be a practical problem. We'll return to this point in the conclusion.

must contain positive instances, although these instances may be very rare. Personal assistant logs often do contain utterances expressing intents which aren't currently implemented, because the functional scope of commercially available personal assistants is continually growing and thus not well understood by all users. This allows intent detectors to be built in advance of functional implementation, at least for intents expressed at the first turn (logs are unlikely to contain subsequent utterances for functionalities that don't yet exist).

A developer begins with a general idea of the classifier they want to build. In our case these will be binary classifiers—in our case, detecting utterances which correspond to a particular intent. For IL we use a tool created in Microsoft Research called ICE, which stands for Interactive Classification and Extraction and is shown in Fig. 1.2.

The developer starts working by searching for data instances using textual search terms based on their domain knowledge. For example, they might issue searches like "calendar" or "my schedule." The searches yield results which the developer then labels. Labels can be positive, negatives, or a special "don't know" label.

After each label is entered, all the labels are randomly divided between a training and test set, and a model is built on the training set, excluding the "don't know" labels. This model is then used in three ways. First, all of the instances labeled so far can be displayed graphically, showing the distribution of scores, giving an overview of performance. Second, when the developer searches for new instances, the model is used to propose labels. This accelerates labeling and also gives an indication of the performance of the model on unseen data. Third, each time the model is rebuilt, it is applied to all of the unlabeled data, which allows the developer to draw samples of unlabeled instances at a particular score. Setting a value near 0.5 will draw instances most perplexing to the classifier, similar to active learning. Setting a high or low value searches for false-positives or false-negatives.

Past work has shown that domain experts (here, developers) can readily provide additional words that can be used as features (Stumpf et al. 2007) and that those words improve machine-learning performance (Stumpf et al. 2009). Therefore, in ICE, the developer can populate a list of individual words or phrases which the developer believes will be important in the domain, like "my calendar," "am i free," and "appointment". Finally, the developer can provide *classes* that pertain to the domain, such as days of the week: Monday, Tuesday, etc. After a feature is edited (by adding or removing a phrase), enabled, or disabled, the model is re-built in a few seconds, and the train and test sets are re-scored. This allows the developer to experiment with different word-based features and immediately see the effects. The developer can also opt to use all observed words/n-grams as features.

As labeling progresses, the developer moves freely between evaluating and improving the model. Evaluation is done by looking at the distribution of scores on labeled instances and the scores assigned to new instances. Improvement is done by adding more labels or editing the features. In addition, in response to the data, the developer may decide to alter the definition of the classifier—i.e., the labeling guidelines they are following—and revise labels accordingly. For example,

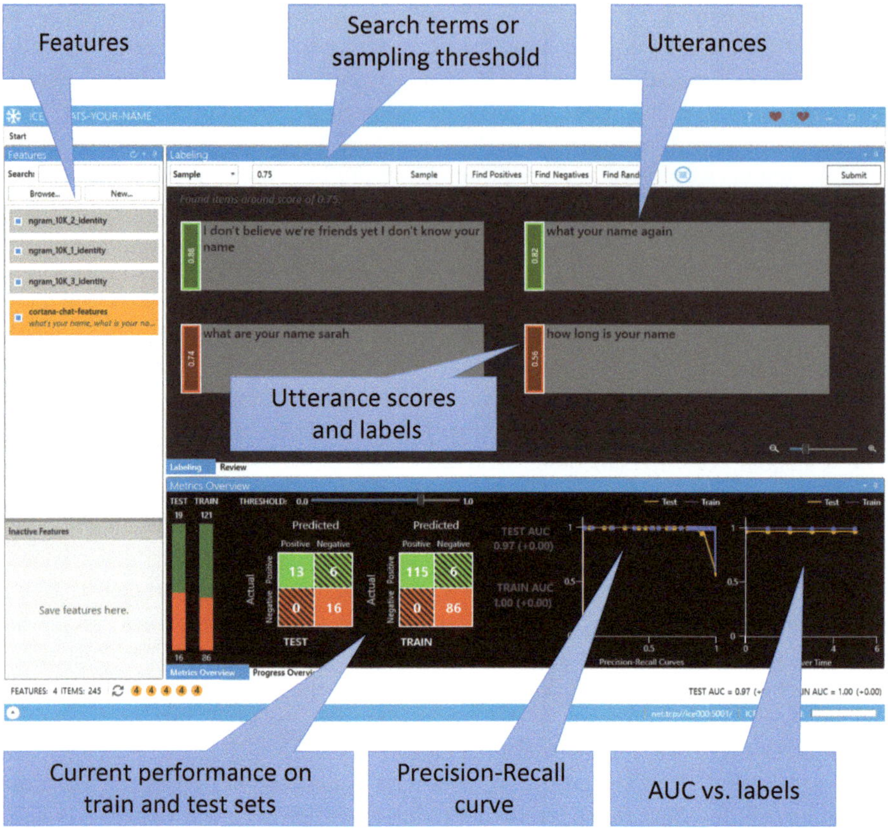

Fig. 1.2 ICE (Interactive classification and extraction): Tool at Microsoft Research for interactive learning (Simard et al. 2014). In the top center, the developer can enter a search term or a score from which to sample. Utterances are shown in the *central panel*. Their scores under the current model are shown to the left of each utterance. The *red* (False) and *green* (True) bars indicate their suggested label by the model, which can be changed by clicking. The performance of the model is shown in the *bottom panel*, including confusion matrices, precision-recall curves, and area under the curve (AUC) as a function of how many labels have been entered. Features can be added and edited on the *left*

the developer may decide that "Show me my calendar" should be a different intent than "Am I free at 3 PM?" when previously they'd been grouped together.

To our knowledge, this work is the first to apply interactive learning to the task of building intent detectors for dialog systems. As compared to the traditional approach which requires about half a dozen staff, IL requires just one person, and that person need not be an expert in machine learning—therefore our hypothesis is that IL will result in substantial time savings over the traditional approach. Active learning addresses the labeling step in the traditional approach—and has been shown to reduce effort at that step—but still requires the same number of roles as the traditional approach.

1.4 Building Intent Detectors with Interactive Learning

As a first test of applying interactive learning to intent detection, we loaded 25M raw utterances from Microsoft Cortana into our IL tool. For typed utterances, the log contained the text entered; for spoken utterances, the log contained the output of the (possibly erroneous) speech recognizer. Utterances likely to contain personal or identifying information were excluded.

We then applied ICE to build three intent detectors in the movies domain:

- MOVIESDIRECTEDBY: the user is requesting to find all movies directed by a named person, for example "What movies did Stanley Kubrick direct?"
- WHODIRECTED: the user is requesting the name of a director for a named movie, for example "Who directed The Matrix?"
- MOVIERUNTIME: the user is requesting the duration of a named movie, for example "How long is Gone with the Wind?"

The first two intents were built to gain familiarity with the IL tool and as a result the effort duration was not carefully logged. The last intent was built under controlled, carefully logged conditions. The developer added bag-of-n-gram features, specific n-grams like "director," "movie," and "who directed," and also a class containing all movies names found in Freebase.[2]

The effort expended (in minutes) is shown in Fig. 1.3. For MOVIERUNTIME, labeling 600 utterances required 90 min. Note that the marginal time per utterance declines sharply: the first 100 utterances required 28 min, whereas the last 100 utterances required 9 min. This illustrates the benefits of interactive learning: early in labeling, the developer is manually searching for utterances to label, the model is unable to suggest labels, and more feature engineering is required; later in labeling, the model can be used to select utterances to label and can often propose accurate labels, and the features are stable so little feature engineering is required.

We then evaluated the performance of all three intent detectors on held-out data. Precision is straightforward to evaluate: the models were run on randomly ordered unseen utterances; the first 150 utterances scored above a threshold were manually labeled. Results are shown in Table 1.1. The precision ranged from 81 to 93 % for the three intents developed.[3]

Unlike precision, recall is infeasible to measure, since each intent appears very infrequently in the data, so accurately estimating recall would require labeling 100 Ks or millions of utterances. As a basic check, 150 unseen utterances were chosen at random, were scored using the MOVIERUNTIME intent detector, and manually labeled. None were found to contain the MOVIERUNTIME intent, and the model scored all below the threshold.

[2]www.freebase.com.

[3]The held-out test set excluded utterances which appeared in the training set, whereas in actual deployment, utterances in the training set may reappear. Therefore, these are conservative estimates which could underestimate performance.

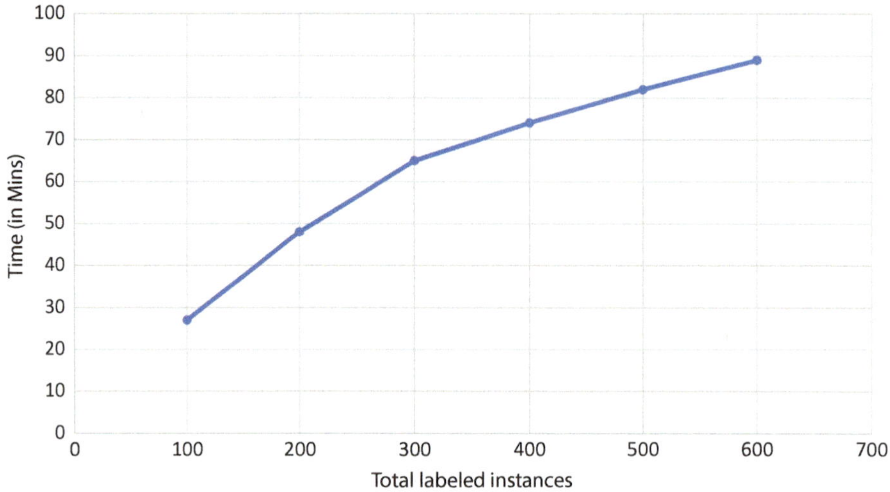

Fig. 1.3 Cumulative effort (time) for building the MOVIERUNTIME intent with interactive learning

Table 1.1 Effort and precision for three binary intent classifiers

Intent	Effort (min)	Number of labels	Test set size	Precision (%)
MOVIESDIRECTEDBY	180[a]	1121	150	93
WHODIRECTED	180[a]	1173	150	89
MOVIERUNTIME	90	600	150	81

[a]Estimate

The false-positives for the MOVIERUNTIME intent were examined by hand. The main cause of errors was that the n-gram "how long is" refers to many types of queries, *and* many words in English are titles of movies, making some instances difficult to classify. For example, in the utterance "How long is Burger King open," the words "Burger" and "King" are both titles of movies, but "Burger King" is not related to movies in this context. This was one consideration in developing a second tool to explore and correct end-to-end operation, described next.

1.5 Improvement Through End-to-End Testing

The results in the previous section showed that it is possible to use interactive learning to quickly build a single intent detector with a good precision, so we next set about building an end-to-end tool for testing and system improvement. Labeling and correcting end-to-end interactions allow developers to view and debug interactions as they will be experienced by users, i.e., developers can decide on intermediate labels for stages in the pipeline based on the response to the user. The design of the tool is to visualize the end-to-end processing done for an utterance. If the developer sees any errors in the pipeline, they can correct the output of any processing stage, and the remainder of the pipeline is immediately rerun using those corrections as the revised input. Once the whole pipeline has been checked and the correct answer is output at the end of the pipeline, the developer can save the labels, which stores labels for every component of the pipeline. The end-to-end tool also allows the developer to type in an utterance, which handles the case where variations of the target intent don't (yet) appear in the logs. This enables the developer to bootstrap intent models when sufficient example utterances do not yet exist. A screenshot of our tool is shown in Fig. 1.4. In the upper left, the developer types in an utterance. The utterance is then scored by all of the available intent detectors—in our case, three—then performs entity identification and calls a function which produces a response.

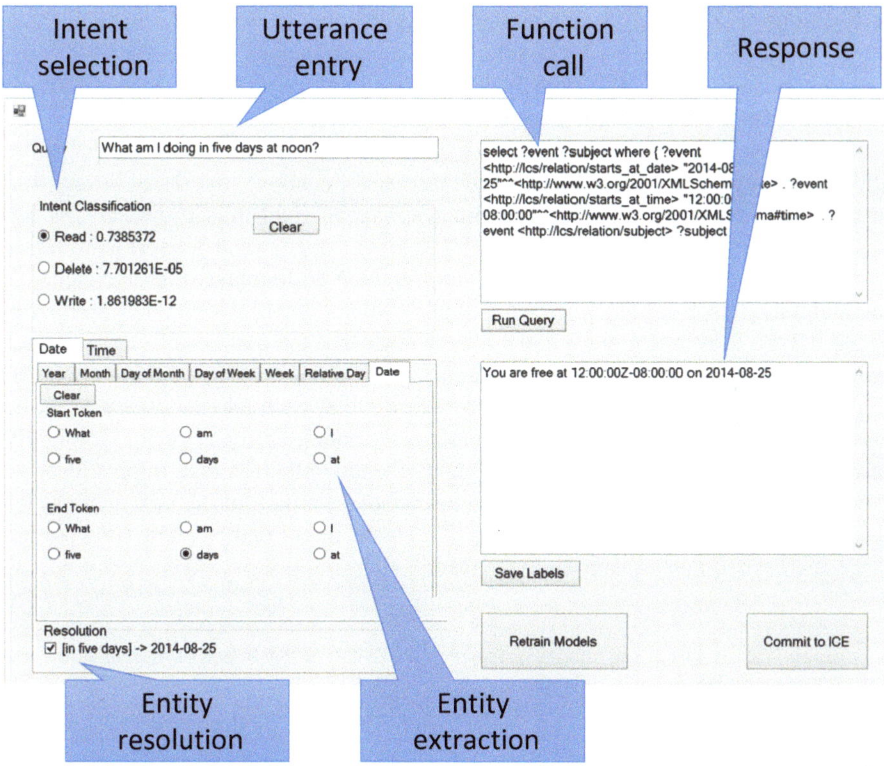

Fig. 1.4 End-to-end labeling tool. The developer enters an utterance in the *upper left*. The intent detectors and their scores are shown below that. The *bottom left* shows entity extraction and resolution. Finally, the resulting function call is shown in the *upper right*, and the system response is shown at *middle right*. Errors detected at any stage of the pipeline can be corrected and are propagated forward; once the whole pipeline produces the correct answer, the developer clicks *Save Labels*, which also clears the display so the developer can enter another utterance. The developer can also *Retrain models* and *Commit to ICE* which stores the new models and labels in the cloud

In this first experiment, we explored the rather limited question of whether it was possible to improve a single intent model using this end-to-end tool. We began with a working calendar dialog system with two intents: READFROMCALENDAR and WRITETOCALENDAR. We then asked a developer to create a new binary intent detector, DELETEFROMCALENDAR, in the ICE IL tool (not in the end-to-end tool). The developer labeled 470 instances in 60 min. On a held-out test set of 3M unlabeled utterances, using a decision threshold of 0.5, the model retrieved 114 utterances, of which 88 % were correct, i.e., the precision was 88 %. In the results below, this model is denoted "IL."

Next, the developer began using the end-to-end tool, with all three intents and two entity extractors. The developer examined 33 utterances in 30 min, and the labels collected were used to update all of the intent and entity models, including the DELETEFROMCALENDAR intent model. The updated model (denoted "IL+E2E") for DELETEFROMCALENDAR was run on the same held-out data used with the baseline (IL) model as above; we adjusted the threshold so that it achieved the same precision as the baseline (IL) model developed in isolation (88 %). At this precision, the IL+E2E model retrieved 134 sentences—an increase of 18 % compared to the baseline model—which was statistically significant at $p = 0.004$ using McNemar's test. These results are summarized in Table 1.2.

This analysis shows that the recall of the model increased for a fixed precision, which in turn means that the F-measure increased. As above, the intent is very rare in the data, so calculation of exact values for recall (and thus F-measure) is not possible.

While it is clear that the end-to-end tool yielded an increase in performance of the intent models, in future work we'd like to evaluate the time efficiency compared to the ICE IL tool in Sects. 1.3 and 1.4. It is clear that the number of utterances per unit time in the end-to-end tool (33 utterances in 30 min) is lower than in the ICE IL tool (470 utterances in 60 min); however we note that each labeled utterance in the end-to-end tool produces labels for *every* intent model, entity extractor, and entity resolver. Therefore it is likely that this end-to-end evaluation also improved these models. We defer further evaluations to future work; rather, in the next section, we continue investigating the performance of a single intent detector in a live, public, large-scale personal assistant dialog system.

Table 1.2 Precision and utterances retrieved for DELETEFROMCALENDAR intent on a test set of 3M unlabeled utterances

Method	Total effort (min)	Number of labels	Precision (%)	Utterances retrieved
IL	60	470	88	114
IL+E2E	90	503	88	134

1.6 Interactive Learning in Production

In this section we report on preliminary results applying this technique to the live Cortana service. Here we apply IL to the social conversation domain, i.e., social utterances directed at the agent, such as "Cortana, where are you from?" or "Are you married?" The designer of this feature—who was not a machine learning expert—developed a binary detector for the COMPLIMENT intent, including utterances like "Cortana you're great" or "That was pretty clever."

The designer spent 135 min developing this classifier, labeling 2254 utterances, approximately half positive and half negative. This classifier was then deployed into the production Cortana platform. A random sample of 144 K utterances from the production platform was then collected. The compliments classifier fired on 1160 of these. These 1160 utterances were manually labeled. Precision was measured to be between 97.7 and 99.9 %, with the variation due to how ambiguous utterances like "that's great" and "very nice" are counted, since these are compliments in some, but not all, contexts. As above, an exact recall measurement is not possible, but in a random sample of 512 utterances, 1 or 2 were compliments (one was ambiguous). This implies an occurrence rate in the data of 0.2–0.4 %; the upper bound of the 95 % confidence interval of the proportion $\frac{1}{512}$ is 1.1 % and $\frac{2}{512}$ is 1.4 %. The classifier fired on 0.8 % of the utterances. These figures cannot be used to compute recall but do suggest that the recall is reasonable, since the fraction of utterances where the (high-precision) classifier fired is about equal to the occurrence rate in a small sample, and not far from the upper limit of the 95 % confidence interval.

By contrast, labeling 2254 utterances selected uniformly at random (rather than selected using the IL tool) is highly unlikely to produce a good model. Assuming 0.5 % of utterances are compliments, then a uniform random sample would result in $0.005 \times 2254 = 11$ positive labels. Learning a good classifier from 11 positive labels and 2243 negative labels would be hopeless.

1.7 Conclusions

Quickly building intent detectors is a major bottleneck for expanding the functionalities of a personal assistant dialog system. In this paper, we have shown how interactive learning can be applied to this problem. Intent detectors have been built in three domains: movies, calendar, and social conversation. Each binary intent detector required 1–3 h to build, and yielded good precision, in the range of 81–99 %, without any obvious problems in recall.

In future work, we will tackle the problem of ensuring comparability of binary classifier scores. Because classifiers produced with interactive learning (or active learning) do not use a training set randomly sampled from the data, their scores are all calibrated differently and are thus not guaranteed to be directly comparable. With a handful of intents, this has not been a practical problem, but in the future there could be 1000s or more binary classifiers running in parallel, with classifiers being added, removed, or changed at any time. The problem of combining many binary classifiers is well studied in the machine learning literature and numerous solutions exist (Allwein et al. 2001; Beygelzimer et al. 2009); what remains to be done is evaluate their applicability to this problem.

In future work we will also consider *structured intents*, where intents can be composed of multiple relations like "Show movies directed by X and starring Y" (Heck et al. 2013). We anticipate binary detectors can be used to detect each relation; the open question is how to compose the detected relations together into a structured intent.

Even so, this paper has illustrated the impact of interactive learning for intent detection. The conventional process for labeling data and building a model has been reduced from weeks to hours, while achieving very high precision. In the process, the number of staff required has been reduced from a half dozen or so to one, and that individual does not require machine learning expertise. We anticipate both of these contributions will help efforts to grow the functionality of personal assistant dialog systems.

Acknowledgements Thanks to Puneet Agrawal for assistance with the Cortana service and to Meg Mitchell, Lihong Li, Sheeraz Ahmad, Andrey Kolobov, and Saleema Amershi for helpful discussions.

References

Allwein EL, Schapire RE, Singer Y (2001) Reducing multiclass to binary: a unifying approach for margin classifiers. J Mach Learn Res **1**:113–141

Beygelzimer A, Langford J, Ravikumar P (2009) Error-correcting tournaments. In: Algorithmic learning theory. Springer, Heidelberg, pp 247–262

Fukubayashi Y, Komatani K, Nakano M, Funakoshi K, Tsujino H, Ogata T, Okuno HG (2008) Rapid prototyping of robust language understanding modules for spoken dialogue systems. In: The Third International Joint Conference on Natural Language Processing (IJCNLP2008)

Glass JR, Weinstein E (2001) Speechbuilder: facilitating spoken dialogue system development. In: EUROSPEECH 2001 Scandinavia, 7th European conference on speech communication and technology, 2nd INTERSPEECH Event, Aalborg, 3–7 September 2001

Haffner P, Tur G, Wright JH (2003) Optimizing SVMs for complex call classification. In: Proceedings of IEEE International Conference on Acoustics, Speech, and Signal Processing, 2003 (ICASSP '03), April 2003

Heck LP, Hakkani-Tür D, Tür G (2013) Leveraging knowledge graphs for web-scale unsupervised semantic parsing. In: Proceedings of INTERSPEECH, Lyon, 25–29 August 2013

Jung S, Lee C, Kim S, Lee GG (2008) Dialogstudio: a workbench for data-driven spoken dialog system development and management. Speech Comm **50**:697–715

Sarikaya R, Hinton G, Ramabhadran B (2011) Deep belief nets for natural language call-routing. In: 2011 IEEE International conference on acoustics, speech and signal processing (ICASSP), pp 5680–5683

Schapire R, Singer Y (2000) Boostexter: a boosting-based system for text categorization. Mach Learn **39**(2–3):135–168

Simard P, Chickering D, Lakshmiratan A, Charles D, Bottou L, Suarez CGJ, Grangier D, Amershi S, Verwey J, Suh J (2014) ICE: enabling non-experts to build models interactively for large-scale lopsided problems. http://arxiv.org/ftp/arxiv/papers/1409/1409.4814.pdf

Stumpf S, Rajaram V et al. (2007) Toward harnessing user feedback for machine learning. In: Proceedings IUI

Stumpf S, Rajaram V et al. (2009) Interacting meaningfully with machine learning systems: three experiments. Int J Hum Comput Stud **67**(8):639–662

Tur G, Mori RD (2011) Spoken language understanding—systems for extracting semantic information from speech. Wiley, New York

Tur G, Hakkani-Tur D, Schapire RE (2005) Combining active and semi-supervised learning for spoken language understanding. Speech Comm **45**(2):171–186

Tur G, Schapire R, Hakkani-Tur D (2003) Active learning for spoken language understanding. **1**:I-276–I-279

Wang YY, Deng L, Acero A (2005) Spoken language understanding. IEEE Signal Process Mag **22**(5):16–31

Chapter 2
News Navigation System Based on Proactive Dialogue Strategy

Koichiro Yoshino and Tatsuya Kawahara

Abstract This paper addresses the concept of information navigation and the system that navigates news articles updated day by day. In the information navigation, the system has a back-end knowledge base and users can access information through a natural interaction. It is composed of several modules that interact with users in different manners. Both the system and the user can take an initiative of dialogue depending on the specification of the user interest. The system allows ambiguous user queries and proactively presents information related to the user interest by tracking the user focus. An experimental result shows that the proposed system based on partially observable Markov decision process and user focus tracking can interact with users effectively by selecting the most appropriate dialogue modules.

Keywords Information navigation • Non-task oriented dialogue

2.1 Introduction

Studies on spoken dialogue systems now enter a new stage. A large number of spoken dialogue systems have been investigated and many systems are now deployed in the real world, most typically as smart phone applications, which interact with a diversity of users. However, a large majority of current applications is based on a specific task description which includes a definite task goal and necessary slots, such as place and date, for the task completion (Hong et al. 1997; Dahl et al. 1994). Users are required to follow these concepts and they need to be aware of the clear task goal according to the system's capability. On the other hand, keyword search systems and question answering systems with a speech interface are also developed for smartphone applications. Such systems can provide answers to a variety of queries from users, but these systems do not conduct dialogue which involves an interaction with users, as they do not incorporate the domain knowledge and dialogue histories

---

K. Yoshino (✉) • T. Kawahara
Academic Center for Computing and Media Studies, Kyoto University, Sakyo, Kyoto 606-8501, Japan
e-mail: yoshino@ar.media.kyoto-u.ac.jp; kawahara@i.kyoto-u.ac.jp

G. Geunbae Lee et al. (eds.), *Natural Language Dialog Systems and Intelligent Assistants*, DOI 10.1007/978-3-319-19291-8_2

(Kupiec 1993; Burke et al. 1997). Moreover, these systems work well only for simple keyword queries and factoid questions, but it is hard to deal with ambiguous user queries or non-factoid questions. These systems assume a clear goal of the user, a unique destination of a dialogue, and the aim of the system is to reach the goal as soon as possible.

However, when users ask something beyond the system's capability of the goal-oriented systems, current system usually replies "I can't answer the question" or turns to the Web search and returns the retrieval list in the display. This kind of dialogue is not a natural interaction since people want to converse with them besides simple commands. A user-friendly conversational system should not reply with "I can't answer the question" even if the system cannot find the result exactly matching the user query (Kawahara 2009). Instead, it should present relevant information according to the user's intention and preference by using domain knowledge and dialogue management that considers the dialogue history. There are several studies towards this direction (Misu and Kawahara 2010; Pan et al. 2012; Heck et al. 2013). This kind of system is realized by information navigation that is addressed in this paper.

2.2 Task of Information Navigation

In human–human dialogue, people usually have topics they plan to talk about, and they progress the dialogue in accordance with the topics (Schegloff and Sacks 1973). Dialogue participants have a role of speaker and listener, and they converse with each other by changing their role of speaker and listener. The proposed system realizes information navigation by taking a role of the speaker who provides information to the user.

An example is shown in Fig. 2.1. First, the speaker offers a new topic and probes the interest of the listener. If the listener shows interest, the speaker describes details of the topic. If the listener asks a specific question, the speaker answers it. On the other hand, if the listener is not interested in the topic, the speaker avoids the details of that topic and changes the topic.

The task of information navigation is designed as non-goal-oriented dialogue according to the above-described manner. The aim of dialogue is to fulfill information demand of the user through an interaction. When the user demands are not always clear, the information navigation system clarifies the user through interactions. The system presents relevant information even if the user request is not necessarily clear and there is no exactly matching result to the user query. Moreover, the system can occasionally present potentially useful information without any explicit request by following the dialogue context.

The task design of information navigation is defined as a selection of information navigation modules. The initiative of dialogue comes and goes between the system and the user because it depends on the specification of the user demand. If the user has a clear demand, the user can ask a specific question that matches to his demand.

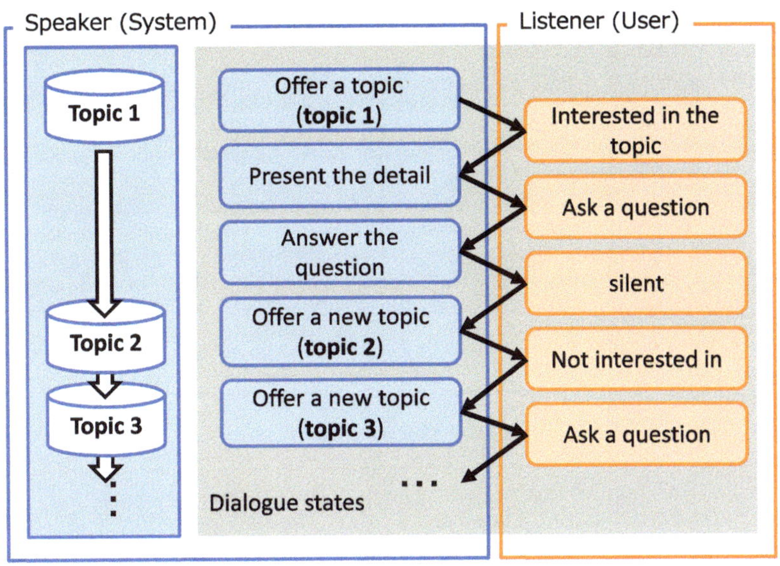

Fig. 2.1 An example of information navigation in human–human conversation

When the user demand is not clear, the system takes an initiative to clarify the user demand by showing candidates that is related to the ambiguous query of the user. This function is achieved by modules that refer to the domain knowledge, the user intention, and the user focus. Here, we define the user focus as "the main piece of information of interest to the user."

In information navigation, the system presents topics that it can talk about, describes the detail of the current topic, or presents topics related to the dialogue history when the system has an initiative. In contrast, the system answers the question of the user, replies to the information demand of the user, or receives a request of changing the topic. The functions of the system modules depend on the kind of information navigation. An example of information navigation modules is shown in Fig. 2.2.

2.3 News Navigation System

We develop a news navigation system that realizes the information navigation described above.

2.3.1 Task of News Navigation

The news navigation system assumes a large number of news articles in raw text as a back-end knowledge source. The knowledge source is limited to the news

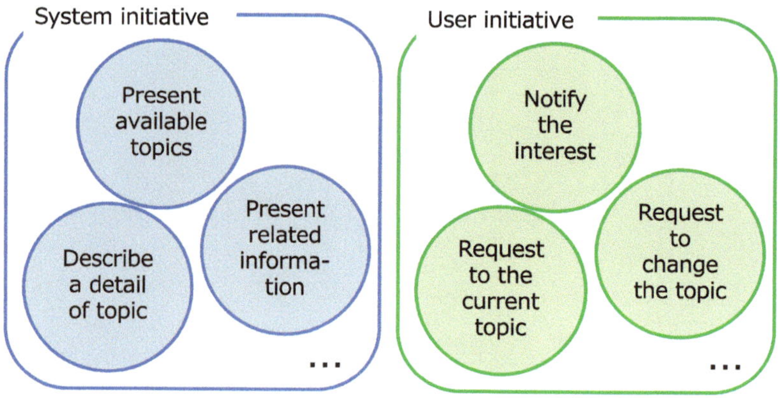

Fig. 2.2 An example of information navigation modules

articles, but the articles are updated day by day. The system navigates this dynamic content by parsing the articles and extracting information from the huge back-end knowledge source. Moreover, it uses a tag of the domain in the news articles to extract the domain knowledge from the text source.

The news navigation system is designed based on the dialogue structure of information navigation depicted in Fig. 2.1. The system gives a briefing on what happened on the day that is written in the articles, and the user can retrieve information through an interaction according to his interests and queries.

2.3.2 System Modules

An overview of the proposed system is illustrated in Fig. 2.3. The system has seven modules, each of which implements a different dialogue act. Each module takes as input a recognized user utterance, an analyzed predicate-argument (P-A) structure, and the detected user focus.

The system begins a dialogue with the "topic presentation (TP)" module, which presents a new topic selected from news articles. It chooses the next module based on the user's response. In this work, it is assumed that each news article corresponds to a single topic, and the system presents a headline of the news in the TP module. If the user shows interest (positive response) in the topic without any specific questions, the system selects the "story telling (ST)" module to give details of the news. In the ST module, the system provides a summary of the news article by using lead sentences. The system can also provide related topics with the "proactive presentation (PP)" module. This module is invoked by the system's initiative; this module is not invoked by any user request. If the user asks a specific question regarding the topic, the system switches to the "question answering (QA)" module to answer the question. This module deals with questions on the presented topic and related topics.

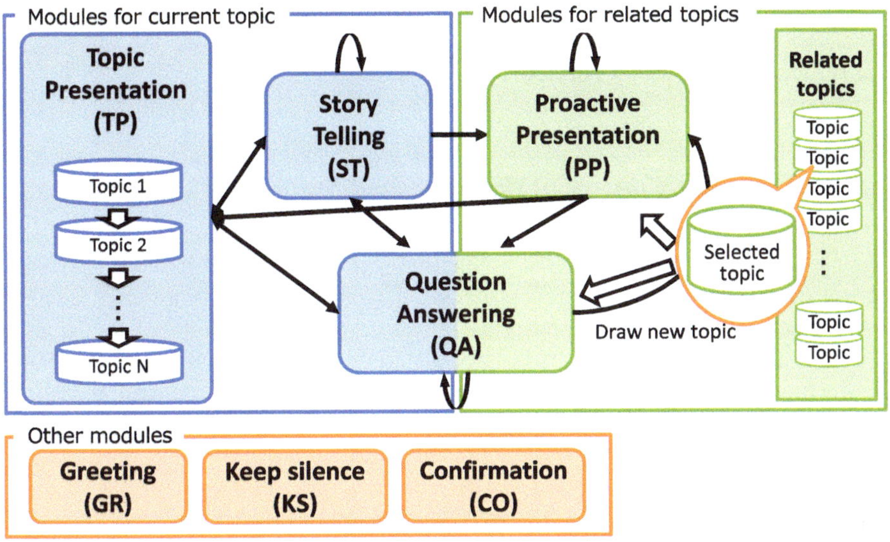

Fig. 2.3 An overview of the information navigation system

The modules of PP and QA are based on a dialogue framework which uses the similarity of the P-A structure between user queries and news articles, and retrieves or recommends the appropriate sentence from the news articles. This method searches for appropriate information from automatically parsed documents by referring to domain knowledge that is automatically extracted from a domain corpus (Yoshino et al. 2011).

Transitions between the modules are allowed as shown in Fig. 2.3. The modules "greeting (GR)," "keep silence (KS)," and "confirmation (CO)" are also prepared. The GR module generates fixed greeting patterns by using regular expression matching. The CO module makes a confirmation if the system does not have certainty about the user query. In terms of dialogue flow, these modules can be called at any time.

The proposed scheme enables the system to answer not only clear requests but also ambiguous requests that do not have any specified goal. The system can respond with flexible matching between the user query and the back-end knowledge source by using the statistical learning result of the semantic P-A structure (Yoshino et al. 2011). As a result, the system has a capability to answer not only factoid questions but also non-factoid questions such as "How was today's Ichiro?" or "How do you feel about the all-star game?" By responding to these questions with some specified news such as "Ichiro hit a home-run" or "28 members are selected for the all-star game," the user can know the outline of the news that he may be interested in, and some more specific questions are invoked.

The dialogue is generated based on the news articles in the knowledge source texts. All modules of the system are automatically trained from the knowledge source, and they are easily portable to different domains.

2.3.3 *Dialogue Control of the Proposed System*

The proposed system is controlled by the dialogue management based on partially observable Markov decision process (POMDP) and conducts information navigation by selecting the most appropriate dialogue module to respond to the user (Yoshino and Kawahara 2014). Markov decision processes (MDPs) and POMDPs are the most successful and now widely used to model and train dialogue managers (Roy et al. 2000; Levin et al. 2000; Williams and Young 2007; Young et al. 2010; Yoshino et al. 2013). These approaches allow us to consider all possible future actions of a dialogue system and thus to obtain a new optimal dialogue strategy which could not be anticipated in conventional hand-crafted dialogue systems.

The conventional scheme for goal-oriented systems assumes that the task and dialogue goal are clearly defined and readily encoded in the reinforcement learning (RL) reward function. This is not true in casual conversation or information navigation addressed in this work.

Some previous work has tackled with this problem. Pan et al. (2012) designed a spoken document retrieval system whose goal is user's information need satisfaction and defined rewards by using the structure of the target document set. This is possible only for well-defined document search problems. The strategy requires a structure of the document set and definition of user demand satisfaction. Shibata et al. (2014) developed a conversational chatting system. It asks users to make evaluation at the end of each dialogue session to define rewards for reinforcement learning. Meguro et al. (2010) proposed a listening dialogue system. In their work, levels of satisfaction were annotated in the logs of dialogue sessions to train a discriminative model. These approaches require costly input from users or developers, who provide evaluation and supervision labels. In the proposed dialogue management, a framework in which reward is defined for the quality of system actions and also for encouraging long interactions is explored, in contrast to the previous approaches. Moreover, user focus is tracked to make appropriate actions, which are more rewarded.

As described in Sect. 2.3.2, the task of information navigation is decoded as a module selection of seven dialogue modules: topic presentation (TP), story telling (ST), question answering (QA), proactive presentation (PP), greeting (GR), keep silence (KS), and confirmation (CO). The dialogue manager selects a module (action decision) based on an input of a user intention. A user intention is encoded as a request to the system; the user intention has six classes and each intention has a corresponding system action.

- *TP*: request to the TP module
- *ST*: request to the ST module

- *QA*: request to the QA module
- *GR*: greeting to the GR module
- *NR*: silence longer than a threshold
- *II*: irrelevant input due to ASR errors or noise

Logistic regression (LR) based dialogue act tagging (Tur et al. 2006) is adopted for the user intention analysis. The existence of the user focus in the utterance is also detected by a discriminative model based on conditional random field (CRF). The system tracks the user focus to select an appropriate action module according to the user interest. The probabilities of the user intention analysis and the user focus detection are used as inputs of belief update of POMDP.

The POMDP updates its belief of the user intention by the recurrence formula

$$b^{t+1}_{s'_j f'_m} = \underbrace{P(o_s^{t+1}, o_f^{t+1} | s'_j, f'_m)}_{\textbf{Obs}} \sum_i \sum_l \underbrace{P(s'_j, f'_m | s_i, f_l, \hat{a}_k)}_{\textbf{Trans}} \, b^t_{s_i f_l}. \tag{2.1}$$

Here, t is a time step and $b^t_{s_i f_l}$ is a belief of the user intention s_i and the user focus f_l. o_s and o_f are observation results of the user intention and the user focus, and $\hat{a}_k$ is the optimal system action selected by the optimal policy function of the POMDP. The POMDP is trained by Q-learning and grid-based value iteration using a user simulator that is constructed from the annotated dialogue data (Yoshino and Kawahara 2014) of news navigation.

Simplified reward for the end of each turn is defined in Table 2.1 to constrain the module selection as an expected behavior. In Table 2.1, + is a positive reward given to appropriate actions, 0 to acceptable actions, and − is a negative reward to inappropriate actions. Here, pairs of a state and its apparently corresponding action, *TP* and TP, *ST* and ST, *QA* and QA, *GR* and GR, and *II* and KS, have positive rewards.

Table 2.1 Rewards in each turn

<table>
<tr><th>State</th><th>Focus</th><th colspan="7">Action a</th></tr>
<tr><th>s</th><th>f</th><th>TP</th><th>ST</th><th>QA</th><th>PP</th><th>GR</th><th>KS</th><th>CO</th></tr>
<tr><td rowspan="2">TP</td><td>0</td><td rowspan="2">+</td><td rowspan="2">−</td><td rowspan="2">−</td><td rowspan="2">−</td><td rowspan="2">−</td><td rowspan="2">−</td><td rowspan="2">0</td></tr>
<tr><td>1</td></tr>
<tr><td rowspan="2">ST</td><td>0</td><td rowspan="2">−</td><td rowspan="2">+</td><td rowspan="2">−</td><td rowspan="2">0</td><td rowspan="2">−</td><td rowspan="2">−</td><td rowspan="2">0</td></tr>
<tr><td>1</td></tr>
<tr><td rowspan="2">QA</td><td>0</td><td rowspan="2">−</td><td>+</td><td rowspan="2">+</td><td>−</td><td rowspan="2">−</td><td rowspan="2">−</td><td rowspan="2">0</td></tr>
<tr><td>1</td><td>−</td><td>+</td></tr>
<tr><td rowspan="2">GR</td><td>0</td><td rowspan="2">−</td><td rowspan="2">−</td><td rowspan="2">−</td><td rowspan="2">−</td><td rowspan="2">+</td><td rowspan="2">−</td><td rowspan="2">0</td></tr>
<tr><td>1</td></tr>
<tr><td rowspan="2">NR</td><td>0</td><td>+</td><td rowspan="2">−</td><td rowspan="2">−</td><td>−</td><td rowspan="2">−</td><td rowspan="2">0</td><td rowspan="2">0</td></tr>
<tr><td>1</td><td>−</td><td>+</td></tr>
<tr><td rowspan="2">II</td><td>0</td><td rowspan="2">−</td><td rowspan="2">−</td><td rowspan="2">−</td><td rowspan="2">−</td><td rowspan="2">−</td><td rowspan="2">+</td><td rowspan="2">0</td></tr>
<tr><td>1</td></tr>
</table>

Other positive rewards are defined for the following reasons. If a user asks a question (*QA*) without a focus (e.g., "What happened on the game?"), the system can continue by story telling (ST). If the system cannot find an answer, it can present relevant information (PP). When the user says nothing (*NR*), the system action should be determined by considering the user focus; present a new topic if the user is not interested in the current topic ($f = 0$), or present an article related to the dialogue history ($f = 1$). Keeping silence (KS) is a safe action to the user silence (*NR*), thus, its reward is 0. However, we give 1 frustration point if the system selects KS in this case because the strategy conflicts with the concept of information navigation. Confirmation (CO) is a safe action to every user input, but it also frustrates the user. Thus, the reward of CO is defined as 0 for every intention, but 2 frustration points are given to the system. If the system selects an inappropriate action (action of $r = -10$), 2 frustration points are given to the system. If the frustration points accumulate more than 10, a large penalty is given to the system and the dialogue is terminated. A large positive reward is given if 20 turns are passed to reward a long continued dialogue.

2.4 Experimental Evaluation

For evaluation of the system, 626 utterances (12 users, 24 dialogues; 2 dialogues with each user) were collected with the proposed dialogue system.

For comparison, we also constructed a rule-based system (=Rule) and a POMDP-based system that does not track the user focus (=POMDP w.o. focus). We evaluated the system performance by the accuracy of action selection. The gold-standard is annotated by two annotators. The agreement for the user states was 0.958 and Cohen's kappa was 0.932. The agreement for the system actions was 0.944 and Cohen's kappa was 0.915. We reprioritized the first annotator who is familiar with the task if the annotation was not agreed.

A breakdown is shown in Table 2.2. The table shows precision (P), recall (R), and F-measure (F) of each intention tag. Here, the results of TP, ST, QA, and PP are presented because the number of KS and GR was very small (#GR = 2, #KS = 4), and CO was not labeled as a correct action. The proposed method outperformed the compared systems for all actions. The proposed method improved the accuracy for topic presentation (TP) and proactive presentation (PP) especially when the user intention was no request (*NR*). The POMDP without the user focus always selected the keep silence (KS) module if the user said nothing (*NR*).

The proposed method also made more effective confirmations (CO) when the SLU result was not correct. It made confirmations (CO) 18 times, and 15 times of them was done when the SLU result was incorrect ($15/18 = 83.3\,\%$). The POMDP without the user focus made only two confirmations, when the detected user intention was correct ($0/2 = 0.0\,\%$).

Table 2.2 Performance of action selection (precision, recall, and F-measure)

	Rule			POMDP w.o. focus			POMDP proposed		
Tag	P	R	F	P	R	F	P	R	F
TP	0.884	0.822	0.852	0.917	0.764	0.834	0.959	0.803	0.874
ST	1.000	0.022	0.043	0.900	0.500	0.643	0.910	0.789	0.845
QA	0.678	0.993	0.806	0.797	0.962	0.872	0.843	0.945	0.891
PP	0.929	0.342	0.500	0.000	0.000	0.000	0.854	0.921	0.886

The proposed method made 35 proactive presentations (PP), and 17 times of them ($17/35 = 48.6\,\%$) invoked new user questions. This result demonstrates that the proposed system encouraged interactions in news navigation.

2.4.1 *Discussion of Trained Policy*

An example dialogue is shown in Fig. 2.4. In the example, the system selects appropriate actions even if the observation likelihood is low. At the 4th turn of Dialogue 1 in this example, the system with the user focus responds with an action of proactive presentation $a = \mathsf{PP}$, but the system without the user focus responds with an action of topic presentation $a = \mathsf{TP}$. At the 2nd turn of Dialogue 2, the user asks a question without a focus. The confidence of $s = QA$ is lowered by the belief update, and the system selects the story telling module $a = \mathsf{ST}$. These examples show that the trained policy reflects the design of information navigation proposed in this paper. It is better to make a proactive presentation when the user is interested in the topic.

2.5 Conclusions

We proposed a new concept of spoken dialogue system that navigates news information. The task of information navigation is along with the direction from conventional task-oriented dialogue systems to general non-task-oriented dialogue systems. In information navigation, users are not forced to accommodate the task goal of the system. Instead, the user can make ambiguous queries. The system provides information that the user wants to know by probing and clarifying the potential demands of the user. The proposed news navigation system can converse with users in a user-friendly manner. It does not respond "I can't answer the question" or turns to the Web search even if it cannot find exact information.

The proposed system is controlled by the dialogue management based on the POMDP that tracks user focus. The system presents relevant information according to the user's interest by tracking the user focus. A POMDP framework is extended to track the user focus to select the appropriate action module. In the experimental evaluations, the system responded with the appropriate action modules according to the user intention and the focus.

Dialogue 1

Dialogue	State
Sys: Hello. Usr: Hello. Tell me today's news. Sys: Hanshin won the game by the game winning hit of Toritani.	$s=TP, P(o\|s)=0.8, b(s=TP)=0.800, f=0$ $a=\mathrm{TP}, b(s=TP f=0)=0.8$
Usr: Did Nohmi come into pitch? Sys: Nohmi was the winning pitcher, by shutting out for seven innings.	$s=QA, P(o\|s)=0.9, b(s=QA)=0.991, f=1$ $a=\mathrm{QA}, b(s=QA, f=1)=1.0$
Usr: Did he get strikeout? Sys: Nohmi struck out Abe on a slider with two out two-on jam.	$s=QA, P(o\|s)=0.9, b(s=QA)=0.994, f=1$ $a=\mathrm{QA}, b(s=QA, f=1)=1.0$
Usr: (pause) Sys: By the way, Nohmi struck out Sakamoto on a 147km/h four-seam in August 29th.	$s=NR, P(o\|s)=1.0, b(s=NR)=1.00, f=1$ $a=\mathrm{PP}, b(s=NR, f=1)=1.0$
Usr: Is four-seam his best pitch? Sys: He struck out Murata with split, his best pitch.	$s=QA, P(o\|s)=0.9, b(s=QA)=0.995, f=1$ $a=\mathrm{QA}, b(s=QA, f=1)=0.9$
Usr: OK, any other news? Sys: Hiroshima stopped losing streak.	$s=TP, P(o\|s)=0.8, b(s=TP)=0.984, f=0$ $a=\mathrm{TP}, b(s=TP, f=0)=1.0$
Usr: Tell me more. Sys: Hiroshima Carp stopped a losing streak at 3 games. ...	$s=ST, P(o\|s)=0.9, b(s=ST)=0.871, f=0$ $a=\mathrm{ST}, b(s=ST, f=0)=0.9$

Dialogue 2

Dialogue	State
Sys: Hello. Usr: Hello. Tell me today's news? Sys: Hanshin won the game by the game winning hit of Toritani.	$s=TP, P(o\|s)=0.8, b(s=TP)=0.800, f=0$ $a=\mathrm{TP}, b(s=TP f=0)=0.8$
Usr: What happened? Sys: Hanshin-tigers won second straight against Yomiuri-giants by the game winning hit of team-leader, Toritani.	$s=QA, P(o\|s)=0.8, b(s=QA)=0.532, f=0$ $a=\mathrm{ST}, b(s=QA, f=0)=0.5$
Usr: When did he hit the game winning hit? ...	$s=QA, P(o\|s)=1.0, b(s=QA)=0.999, f=1$

Fig. 2.4 A dialogue example

References

Burke RD, Hammond KJ, Kulyukin V, Lytinen SL, Tomuro N, Schoenberg S (1997) Question answering from frequently asked question files: experiences with the faq finder system. AI Mag 18(2):57–66

Dahl DA, Bates M, Brown M, Fisher W, Hunicke-Smith K, Pallett D, Pao C, Rudnicky A, Shriberg E (1994) Expanding the scope of the ATIS task: the ATIS-3 corpus. In: Proceedings of the workshop on human language technology, HLT 1994, pp 43–48

Heck L, Hakkani-Tur D, Chinthakunta M, Tur G, Iyer R, Parthasarathy P, Stifelman L, Shriberg E, Fidler A (2013) Multimodal conversational search and browse. In: IEEE workshop on speech, language and audio in multimedia, August 2013

Hong L, Muraki S, Kaufman A, Bartz D, He T (1997) Virtual voyage: interactive navigation in the human colon. In: Proceedings of annual conference on computer graphics and interactive techniques, SIGGRAPH 1997, pp 27–34

Kawahara T (2009) New perspectives on spoken language understanding: does machine need to fully understand speech? In: IEEE workshop on automatic speech recognition & understanding, ASRU 2009, pp 46–50

Kupiec J (1993) Murax: a robust linguistic approach for question answering using an on-line encyclopedia. In: Proceedings of annual international ACM SIGIR conference on research and development in information retrieval, SIGIR 1993, pp 181–190

Levin E, Pieraccini R, Eckert W (2000) A stochastic model of human-machine interaction for learning dialog strategies. IEEE Trans Audio Speech Lang Process 8(1):11–23

Meguro T, Higashinaka R, Minami Y, Dohsaka K (2010) Controlling listening-oriented dialogue using partially observable Markov decision processes. In: Proceedings of international conference on computational linguistics, COLING 2010, pp 761–769

Misu T, Kawahara T (2010) Bayes risk-based dialogue management for document retrieval system with speech interface. Speech Commun 52(1):61–71

Pan Y-C, Lee H-Y, Lee LS (2012) Interactive spoken document retrieval with suggested key terms ranked by a Markov decision process. IEEE Trans Audio Speech Lang Process 20(2):632–645

Roy N, Pineau J, Thrun S (2000) Spoken dialogue management using probabilistic reasoning. In: Proceedings of annual meeting on association for computational linguistics, ACL 2000, pp 93–100

Schegloff EA, Sacks H (1973) Opening up closings. Semiotica 8(4):289–327

Shibata T, Egashira Y, Kurohashi S (2014) Chat-like conversational system based on selection of reply generating module with reinforcement learning. In: Proceedings of international workshop series on spoken dialog systems, IWSDS 2014, pp 124–129

Tur G, Guz U, Hakkani-Tur D (2006) Model adaptation for dialog act tagging. In: Proceedings of IEEE workshop on spoken language technology, IWSLT 2006, pp 94–97

Williams JD, Young S (2007) Partially observable Markov decision processes for spoken dialog systems. Comput Speech Lang 21(2):393–422

Yoshino K, Kawahara T (2014) Information navigation system based on POMDP that tracks user focus. In: Proceedings of annual SIGdial meeting on discourse and dialogue, SIGDIAL 2014, Philadelphia, PA, pp 32–40, June 2014

Yoshino K, Mori S, Kawahara T (2011) Spoken dialogue system based on information extraction using similarity of predicate argument structures. In: Proceedings of annual SIGdial meeting on discourse and dialogue, SIGDIAL 2011, Portland, OR, pp 59–66, June 2011

Yoshino K, Watanabe S, Le Roux J, Hershey JR (2013) Statistical dialogue management using intention dependency graph. In: Proceedings of international joint conference on natural language processing, IJCNLP 2013, Nagoya, pp 962–966, October 2013

Young S, Gašić M, Keizer S, Mairesse F, Schatzmann J, Thomson B, Yu K (2010) The hidden information state model: a practical framework for POMDP-based spoken dialogue management. Comput Speech Lang 24(2):150–174

Chapter 3
Evaluation of Machine-Led Error Recovery Strategies for Domain Switches in a Spoken Dialog System

Sven Reichel, Ute Ehrlich, André Berton, and Michael Weber

Abstract Spoken dialog systems which include multiple domains or many applications set high requirements for natural language understanding. As the functionality in such systems increases, recognition errors and ambiguous interpretations are likely to occur. However, switching a domain or application by accident reduces user satisfaction and task success rate enormously. Therefore, efficient error recovery strategies need to be applied. In an online study, we evaluated three different machine-led error recovery strategies for in-car infotainment systems. They are varied first in terms of modality (visual and speech) and second in using contextual information. By comparing the strategies, we figured out that asking novice users an open question does not work and they prefer to select the domain from a list of alternatives. This list needs to be minimized concerning number of items and has to contain the requested one. A trade-off between list length and confidence has to be made, based on partial interpreted user utterances and correct predictions of follow-up domains. Furthermore, a choice out of two items requires a graphical visualization, whereby a list performs good with an acoustic presentation and does not need visual elements.

Keywords Error recovery • In-car infotainment • Multi-domain • Spoken dialog system • User study

3.1 Introduction

More and more people are "always on" due to the success of smartphones or other web-enabled devices. The power of these devices increases every year and people use them more than ever. A study by the Nielsen Company shows that app usage in

S. Reichel (✉) • U. Ehrlich • A. Berton
Daimler AG, Ulm, Germany
e-mail: sven.reichel@daimler.com; ute.ehrlich@daimler.com; andre.berton@daimler.com

M. Weber
Institute of Media Informatics, Ulm University, Ulm, Germany
e-mail: michael.weber@uni-ulm.de

G. Geunbae Lee et al. (eds.), *Natural Language Dialog Systems and Intelligent Assistants*, DOI 10.1007/978-3-319-19291-8_3

the U.S. rose about 65 % from 2012 to 2013 (The Nielsen Company 2014). However, the classical app interaction schema, such as opening an app, interacting with it, and switching to another one, is altered by personal assistants (e.g. Apple's Siri[1] or Microsoft Cortana[2]), which are able to recognize and execute user intentions from various domains. For instance, they can search for restaurants, call a selected one to reserve a table, navigate you there, and additionally they will tell you Point-of-Interests on the way, all without switching the app. This is possible because they rely heavily on user-initiated natural speech interaction, which enables users to say whatever they like.

However, "building a dialog management system for the processing of dynamic multi-domain dialogs is difficult" as Lee et al. (2009) stated. One crucial point is to identify the domain of interest correctly to process the user's request. This is not an easy task to do, as multi-domain or open-domain spoken dialog systems (SDSs) require large language models, which decrease the speech recognition accuracy and language understanding (Carstensen et al. 2010). Thus, an SDS can never be completely sure, whether the user really intends a domain switch or not.

Processing the domain switch correctly within a multi-domain SDS is crucial to user satisfaction and task success. On the one hand, switching a domain by accident will require the user to correct or even restart the dialog. On the other hand, not recognizing a domain switch may prevent users from reaching their task goal. While these are more or less user satisfaction issues on a smartphone, for in-car systems they affect the driver's safety seriously. As we have shown in Reichel et al. (2014), a non-expected infotainment system behavior results in an increase of driver distraction. As a result, in-car systems need to pay special attention to domain switching and out-of-domain utterances.

Considering this fact, what can in-car systems do if the confidence score of a potential domain switch is low? In this paper, we present different error recovery or clarification strategies, which were evaluated with an online study concerning task success and usability. In Sect. 3.2 we provide an overview of existing approaches before presenting our strategies in Sect. 3.3. Section 3.4 describes the study's setup to evaluate our strategies. Results are presented and discussed in Sect. 3.5, before we conclude in Sect. 3.6.

3.2 Error Recovery Strategies in Multi-domain SDSs

As Steve Young pointed out in his keynote at SigDial 2014 (Young 2014), current SDSs are designed to operate in specific domains, but for accessing web-based information and services, open-domain conversational SDSs are needed. In a previous explorative study (Reichel et al. 2014), we figured out that users do not want to switch between various applications explicitly, instead natural switching

---

[1] https://www.apple.com/ios/siri/, online accessed 2015/03/04.

[2] http://www.windowsphone.com/en-us/features-8-1#Cortana, online accessed 2015/03/04.

between different services should be possible. SmartKom (Reithinger et al. 2003) was one of the first SDSs to provide a multimodal interface (Smartakus) for accessing 14 different applications. It is built upon a closed-world ontology and it only understands what is modeled. Recognition errors, or user utterances which are out of domain, are tried to be corrected on a technical level (e.g., query relaxation). Various other technical approaches exist to process domain switches and out-of-domain utterances correctly (e.g., Nakano et al. 2011; Robichaud et al. 2014). However, they fail for domain ambiguous utterances and even in SDSs using open-world knowledge bases for robust task prediction, situations may occur in which user utterances are ambiguous and an explicit clarification by the user is needed (Pappu and Rudnicky 2013).

Bohus and Rudnicky (2005) analyzed various recovery strategies and identified the "move on" (ignoring the error first and correcting it later on) and "help the user" (providing help messages with sample responses) strategy as good approaches for explicit clarification. However, these are generic clarification strategies, which are used less often in human–human dialogs. Humans prefer context-aware, targeted clarifications to resolve automatic speech recognition (ASR) errors (Stoyanchev et al. 2014). Skantze's approach (Skantze 2007) also relies heavily on dialog context and partially interpreted user utterances to handle errors in different modules of SDSs. These approaches do not consider domain switches, which often face problems in terms of ambiguities and out-of-domain utterances, thus non-understanding of the complete utterance.

An overview of different error-recovery strategies for multimodal and pervasive systems is provided by Bourguet (2011). She classifies all strategies according to actor, modality, and purpose. In our work, the purpose is always to make users clarify the domain their utterance refers to (error correction). Concerning actor and modality, variations of the strategies are developed (see Sect. 3.3).

3.3 Helping the User During a Domain Switch

In a previous experiment (Reichel et al. 2014), we analyzed successful and non-successful domain switches during a driving situation. An error recovery strategy, in which the system takes the initiative and tells users what they can say [Notify and YouCanSay strategy (Bohus and Rudnicky 2005)], was compared to them. Concerning task success, this strategy was only 3.3 % worse than the successful domain switch, however, it's usability scores and the driver's distraction tended towards the nonsuccessful domain switch. The prompt to tell people what to say was too long and narrative.

Based on these results, three recovery strategies and a reference system were developed to handle uncertainty of domain switches by clarification requests (cf. Appendix):

Reference (REF): An optimal system understands a user's request and executes the desired action. However, as a false domain switch and the requested action would result in severe consequences (e.g., booking a hotel), an explicit confirmation question is always asked. As each participant rates a dialog system on different aspects, we included the reference system to consider these variances. This enables us to compare our strategies with an optimal system.

Ask the User (AU): Asking a user to clarify her intention is always possible for an SDS. Questions can be put in a directed (e.g. "Do you mean Hotel or Facebook") or open-ended (e.g. "Which application are you addressing with your request?") prompt (Jacko 2012). The AU strategy uses open-ended prompts, which do not restrict users to certain keywords. However, users need to anticipate or know what the system is able to understand (the system's applications).

Domain Choice (DC): Directed dialogs do not require any knowledge of the user as they make clear what the system understands (Zoltan-Ford 1991). By having only a limited number of alternatives, the system is able to provide them to the user at once. We propose a choice out of two alternatives. However, in multi-domain SDSs, there might be dialog states in which more than two possible domain switches are likely. This increases the risk to present only wrong alternatives, which slows down the error correction process (Suhm et al. 2001).

Domain List Selection (DLS): If the number of alternatives increases, a list can be presented. While lengthening the prompt, this will reduce the risk to present only wrong alternatives. Users are able to interrupt the prompt by using barge-in after they heard the keyword, which will lead to their task goal. We explained the barge-in and facilitated it by using a short pause after each keyword.

These three dialog strategies enable an SDS to handle cross-domain utterances efficiently. First, they can be used to clarify domain switches in case of low confidence scores. Second, out-of-domain utterances can be classified by the user to the corresponding domain and can be reinterpreted with the right language models.

3.3.1 *Variations of the Dialog Strategies*

The success of the dialog strategies may depend on the kind of presentation and use of contextual information. In-car infotainment systems are normally equipped with a display and speakers, so multimodal output can be used. Visual output requires the driver to look at the display and thus increases gaze-based distraction (Barón and Green 2006; Hofmann et al. 2014). The REF and AU strategies do not require to present any visual information to the user during a domain switch. However, the most probable follow-up domains in the DC and DLS strategies can be presented using both available modalities. As Suhm et al. (2001) showed multimodal error correction strategies are more accurate than unimodal ones. Considering this fact, three different kinds of presentation for each error recovery strategy were developed:

GUI focused (Gui): The idea behind this implementation is to keep the prompts as short as possible. A generic question (e.g. "Say an application name or line number.") is asked and the alternatives are only displayed on the screen.

Speech focused (S): This variant does not present any dynamic information on screen. Alternatives are only read out as presented in the Appendix.

GUI & Speech (GS): Multimodal output is used to present alternatives on screen and reading them out simultaneously. After selecting an alternative in the DC strategy, it will be highlighted to confirm the selection. The list of alternatives in the DLS strategy is scrolled dynamically, whereby highlighting and reading out is synchronized.

For presenting the alternatives according to the DC and DLS strategy, the system has to decide in which order they appear. As applications are more or less static, it could present them in a fixed order. However, humans usually do not use such a generic clarification strategy and react context-aware (Stoyanchev et al. 2014). Therefore, we compare two systems, one **with context (withCtx)** and another one **without context (withoutCtx)**. The system with context predicts the most probable follow-up application based on dialog state and user utterance. This application is presented within the two alternatives of the DC strategy and it is added to the top of the DLS list. Without context, the system does present two wrong alternatives for the DC strategy and it inserts the correct application further down of the list, so that scrolling is necessary.

3.3.2 *Hypotheses*

The different variants of our recovery strategies are evaluated concerning usability and task success. It can be assumed that differences exist between strategies, context, and kind of presentation. Table 3.1 shows the hypotheses. An interesting part is the performance of our error recovery strategies. We assume that significant differences exist between the three strategies and the reference system will perform best (H1). The conditions which consider the context are expected to perform better than the ones without context, as users will reach their task goal more efficiently (H2). Concerning the kind of presentation, no significant differences are assumed in task success because all variants contain the same information. However, users will have preferences for certain kinds of presentation (H3).

Table 3.1 Hypotheses to evaluate (= no sig. diff.; ≠ sig. diff.; > sig. better than)

Hypothesis	Dimension	Task success	Usability
H1	Strategies	$REF > AU \neq DC \neq DLS$	$REF > AU \neq DC \neq DLS$
H2.1	Context	$DC_{withCtx} > DC_{withoutCtx}$	$DC_{withCtx} > DC_{withoutCtx}$
H2.2	Context	$DLS_{withCtx} > DLS_{withoutCtx}$	$DLS_{withCtx} > DLS_{withoutCtx}$
H3	Presentation	$G = S = GS$	$G \neq S \neq GS$

3.4 Evaluation of Dialog Strategies with an Online Study

The error recovery strategies presented in Sect. 3.3 are evaluated with an online user study. This kind of evaluation method allows access to a large number of people in a short time. However, two drawbacks of online studies are missing contextual situation and different interpretation of questions (Lazar et al. 2010). For now, we neglect the driving situation in favor of many participants and focus on usability as well as task success. In the next step, the best strategies will be implemented and evaluated in a driving simulator. When we designed the GUI, we followed the standardized AAM guidelines (Driver Focus-Telematics Working Group 2006), which will prevent major driver distraction and prepares the integration into a car's infotainment system. The problem with different interpretations of questions is addressed by using only validated questionnaires.

3.4.1 User Tasks

In a user study, it is crucial to set real tasks for users, as with artificial ones they cannot put themselves into the situation. By using a calendar entry for dialog context (see Fig. 3.1), multiple cross-domain tasks can be imagined. The different semantic values, namely title, date, location, participant, and description, can be used to trigger other tasks. Table 3.2 shows the tasks we used in this study. They are classified into information seeking (inf) and action (act) tasks. This is based on Kellar et al.'s classification schema (Kellar et al. 2006) whereby information exchange and maintenance are grouped together and named action tasks, as they initiate an action.

While tasks occur in real life naturally, in a study users have to be briefed to know their task. This can be achieved through a variety of means. Bernsen and Dybkjaer (1997) suggest written instructions or graphically depicted scenarios. However, written instructions prime users to these words and no variances in utterances will be collected. Therefore, we use graphically depicted scenarios.

Fig. 3.1 Dialog context for starting a cross-domain task: participants see a calendar entry showing a concert of Elton John at New York on August 21st, 2014. Alexandra will be there too and a note identifies Elton John's new album. From this dialog state, multiple domain changes are possible (cf. Table 3.2)

Table 3.2 Cross-domain user tasks

Task	Semantic value	New domain	Example user utterance	Type
T1	Date	Hotel	"Book a hotel for this concert"	act
T2	Date, Location	Weather	"Tell me the weather"	act
T3	Location	Knowledge	"What is the Carnegie Hall?"	inf
T4	Participant	Phone	"Call Alexandra to cancel the appointment"	act
T5	Description	Music	"Play the new album on the Internet radio"	act
T6	Location	Navigation	"Navigate me there"	act
T7	Title	Facebook	"Share this appointment on Facebook"	act
T8	Location	Knowledge	"When was this location established?"	inf
T9	Title	Knowledge	"When was the artist born?"	inf

Table 3.3 Each participant evaluates one variant

Variant	Presentation	Context	Dialog strategies
Gui	GUI focused	–	REF, AU, DC, DLS
GS_withoutCtx	GUI & Speech	Without	REF, AU, DC, DLS
GS_withCtx	GUI & Speech	With	REF, AU, DC, DLS
S_withoutCtx	Speech focused	Without	REF, AU, DC, DLS
S_withCtx	Speech focused	With	REF, AU, DC, DLS

3.4.2 Design of the User Study

As described in Sect. 3.3, three recovery strategies and one reference system are developed. These are rated and compared with each other by each participant. There are three variations in terms of presentation and two which are affected by context. By combining the context with each presentation, six variants would emerge. However, varying context in Gui is not reasonable, because of two issues in DLS (cf. Appendix (d)). First, scrolling the list of alternatives would require an additional user interaction step and thus would disadvantage the Gui condition. Second, it is not clear where to present the requested alternative in the list (top, middle, or bottom), because people may start to read at different screen regions. As a result, Gui is implemented in one variant, positioning the requested alternative at different positions. If people are able to compare different variants (Gui, Speech, Gui&Speech), it is likely that they will prefer the multimodal presentation. However, as the system should be implemented within a car, visual distraction is a matter. So we want to figure out whether people need a visual representation or acoustic would be enough. Thus each participant evaluates four dialog strategies in one variant (cf. Table 3.3).

The strategies are evaluated concerning task success and usability. For task success, the user utterances after a system prompt are manually annotated, regarding whether the participant was able to respond correctly or not. Correctly means, an

SDS would be able to maintain the dialog flow towards task success. Usability is rated with some questions of the subjective assessment of speech system interfaces (SASSI) questionnaire (Hone and Graham 2000). Questions concerning the dimensions Likability, Annoyance, and System Response Accuracy are asked. As participants only rate one system utterance, asking questions concerning the general system performance is not feasible. In addition, three questions from ITU-T Rec. P.851 (International Telecommunication Union ITU) are asked: help (7.3 Q4), concentration (7.2 Q6), and overall impression. Answers are provided with a 7-point Likert scale from strong disagreement (−3) to strong agreement (+3). The six dimensions are averaged to one usability score.

3.4.3 *Procedure of the Experiment*

Five variants are required and were implemented with the online tool LimeSurvey.[3] As each participant only takes part in one variant, five groups of participants are needed. However, Hempel (2006) observed that users' age, gender, and technical experience influence the usability rating and task success of telephone-based SDSs. This means the five groups should have equal populations concerning these attributes. Therefore, we use Hoare et al.'s adaptive random sampling method with stratification (Hoare et al. 2013) to assign participants to a group after they submitted their age, gender, and experience. The link to the study was published via different channels, such as email, mailing lists, personal invitation, flyer, poster, and Facebook.

At the beginning, participants are asked to provide personal data in a questionnaire. After that, the experiment consists of two parts: in the first one participants provide utterances by themselves and in the second one they see videos of sample interactions. Part one requires participants to complete a task with each strategy (strategies are sorted due to learning effects: REF, AU, DC, and DLS) and rate it afterwards. We record the participants' utterances, whereby the system responses are pre-recorded videos (see Appendix for end-to-end sample dialogs). The pre-recorded videos can only be played once, as we want to analyze task success and by repeating the system responses this result would be biased. In addition, barge-in is possible, however, resumption is permitted. After completing the four tasks, participants compare them on a 7-point Likert scale. In the second part of the study, the questionnaires and comparisons are the same as in the first one, but participants judge sample interactions in third person view. This part is randomized, as participants do not need to answer on the questions by themselves, so it does not matter when they see the correct answer for AU in the list of DLS.

[3]http://www.limesurvey.org, online accessed 2014/09/18.

3.5 Results and Discussion

In the following, evaluation results of the four dialog strategies are shown. We analyzed data from 99 participants (71m/28f), with average age of 30.4 years ($SD = 9.7$). They have a medium experience with SDSs (6-Likert Scale, $M = 3.3$, $SD = 1.37$), but in general they are technical affine (5-Likert Scale, $M = 3.99$, $SD = 0.68$). Eight participants had problems with their microphone (8m/1f) and five aborted after the first part (2m/3f). Nearly all of the tasks were understood correctly by the participants (95 %), which confirms our approach with visual task descriptions.

In terms of usability, we assessed four usability scores: (1) rating of each strategy, (2) comparison of the four strategies, (3) rating of each sample interaction, and (4) comparison of the sample interactions. We compared them for each dialog strategy with a repeated measures ANOVA test. No significant differences were found between (1) and (2). However, the AU strategy is rated better in the sample interaction videos than in the interactive part, $F(1, 81) = 14.07, p < 0.001, \eta^2 = 0.148$ (Helmert Contrast). For DLS this is similar, $F(1, 81) = 5.82, p = 0.018, \eta^2 = 0.067$ (Helmert Contrast). As (1) and (2) are ratings from first person view which are based on real interactions, we use (1) for further comparisons.

3.5.1 *Evaluation of the Dialog Strategies (Hypothesis 1)*

The strategies are compared in terms of usability and task success. Figure 3.2a shows usability scores of the strategies from (1), which differ significantly, $F(3, 258) = 113.46, p < 0.001, \eta^2 = 0.569$. REF is rated best and AU worst. DLS and DC are in between, however DC depends heavily on context (cf. Sect. 3.5.2). Concerning task success (see Fig. 3.2b), REF and DLS are very high and users nearly always reach their goal. In *REF_GS_withoutCtx* some users neglected the

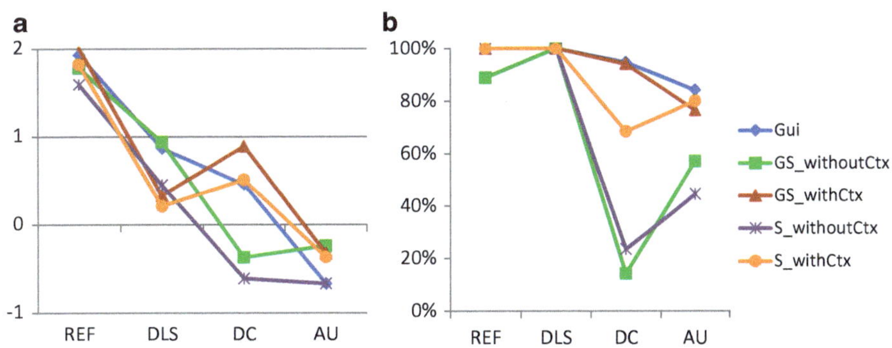

Fig. 3.2 Results of the interactive part. (**a**) Usability rating. (**b**) Task success

explicit confirmation thus task success is lower than in other variants. As with usability, DC depends on context. It can be seen that open questions, such as in AU, do not work properly with novice users. However, AU's usability score of the sample interaction (0.42) shows significant differences compared to the real interaction part (−0.44), $t(85) = -4.74, p < 0.001$. This leads to the assumption that if users know what application answers the request they will rate the AU strategy better.

3.5.2 Using Contextual Information (Hypothesis 2)

Analyzing the results concerning contextual information shows importance of context. In DLS and DC the context affects the order of applications, whereby in AU the kind of task is varied (action and information retrieval task). In DLS no significant differences could be identified concerning task success or usability. However, the usability of the without context conditions is rated slightly better than with context. This might be due to the fact that only 37 % of participants used barge-in. The others heard the list of applications till the end and had to remember the requested one. As seen in Fig. 3.2, DC depends heavily on context. By showing the requested application, the usability score is on the same level as DLS, otherwise it is worse, $t(70) = 4.25, p < 0.001$ (GS and S combined). The effect on task success is even worse, only around 20 % of the participants reached their goal. In AU there is no significant difference concerning usability. However, task success identified problems in terms of identification of the right application for information retrieval tasks. High variances in the requested application can be seen, such as "Websearch," "Browser," "Wikipedia," or "Google." AU and DC strategy may perform better with expert users, but for novice users their success depends on task type and context.

3.5.3 Presentation with Different Modalities (Hypothesis 3)

As REF and AU do not require to present any visual information to the user, only DLS and DC are compared. We hypothesized that task success does not depend on the kind of presentation, but usability scores do. As context affects the results (see Sect. 3.5.2) we compare conditions with the same contexts (*GS_withCtx* vs. *S_withCtx* and *GS_withoutCtx* vs. *S_withoutCtx*). The usability rating (Fig. 3.2a) shows that in DLS and DC the GUI & Speech variant (GS) is slightly better than the speech focused (S) variant, however, none of these differences are significant (*withCtx*: $t(72) = 0.86, p = 0.40$; *withoutCtx*: $t(65) = 1.10, p = 0.28$). Concerning task success, a difference can be seen between *S_withCtx* and *GS_withCtx* for DC only. If users were asked "Do you want A or B" they often responded "yes," which cannot be processed by any SDS correctly. A visual representation makes the selection clearer and leads to improvements of task success.

3.6 Conclusions

In this work we compared different error recovery strategies for domain switches in SDSs. Obviously, a successful domain switch performs best in terms of usability and task success. However, in case of uncertainty about a domain switch, an SDS should be able to ask the user for clarification. Our results show that an open question, such as "Which application are you addressing with your request?", does not work for novice users (especially information retrieval tasks are critical). We compared this approach with two recovery strategies in direct prompting style: first a choice out of two alternatives and second, a list selection out of nine items. The results show that the domain choice is a reasonable approach, if the requested application is within the presented alternatives. The domain list allows users to select the right application easily and achieves good usability scores. So far, the dialog strategies are only evaluated with novice users and not in a real driving situation. Expert users, who have learned the interaction schema with machine-led correction strategies, might react appropriately on open questions and thus would be able to interact efficiently. Furthermore, in the car the domain choice might perform better, as duration is a matter. Each second the driver is occupied by the SDS, she might be distracted from the road. In our sample task the domain choice took 6 s, whereby the list took 20 s. However, domain choice requires a graphical visualization, whereby the list performs good with an acoustic presentation and does not need visual elements.

As a result, each strategy has advantages and disadvantages. Therefore, in the future an adaptive approach has to be considered which adapts the error recovery strategy based on the user (novice or expert) and number of predicted follow-up domains. If only two domains are likely a choice can be used, otherwise a selection list will be better. An intelligent solution has to be developed to limit the number of follow-up domains based on the current dialog state and partial interpreted user utterance. Based on these, an adaptive strategy can be implemented in a car's infotainment system and can be evaluated in a driving situation.

Acknowledgements The work presented here was funded by GetHomeSafe (EU 7th Framework STREP 288667). We would like to thank the EC for funding the GetHomeSafe project.

Appendix

Graphical and speech dialog implementation of the four error recovery dialog strategies (speech dialogs translated from German):

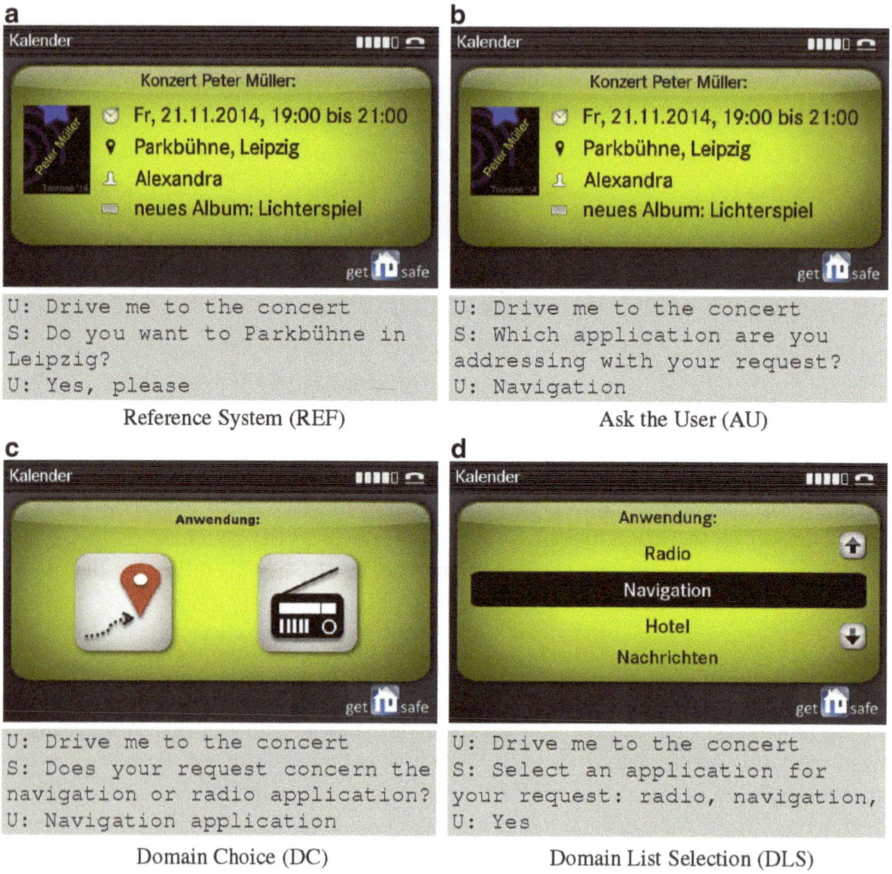

References

Barón A, Green P (2006) Safety and usability of speech interfaces for in-vehicle tasks while driving: a brief literature review. Technical report, University of Michigan TRI

Bernsen NO, Dybkjaer L (1997) Designing interactive speech systems: from first ideas to user testing, 1st edn. Springer, New York

Bohus D, Rudnicky AI (2005) Sorry, i didn't catch that! an investigation of non-understanding errors and recovery strategies. In: Proceedings of SIGdial, Lisbon

Bourguet ML (2011) Uncertainty and error handling in pervasive computing: a user's perspective. In: Ubiquitous computing, Chap 3, Babkin, Eduard

Carstensen KU, Ebert C, Ebert C, Jekat S, Klabunde R, Langer H (2010) Computerlinguistik und Sprachtechnologie. Spektrum, Akad. Verl.

Driver Focus-Telematics Working Group (2006) Statement of principles, criteria and verification procedures on driver interactions with advanced in-vehicle information and communication systems. Alliance of automotive manufacturers

Hempel T (2006) Usability of telephone-based speech dialog systems as experienced by user groups of different age and background. In: 2nd ISCA/DEGA tutorial and research workshop on perceptual quality of systems, Bonn

Hoare Z, Whitaker C, Whitaker R (2013) Introduction to a generalized method for adaptive randomization in trials. Trials 14(1):19
Hofmann H, Tobisch V, Ehrlich U, Berton A, Mahr A (2014) Comparison of speech-based in-car hmi concepts in a driving simulation study. In: Proceedings of IUI, Haifa
Hone KS, Graham R (2000) Towards a tool for the subjective assessment of speech system interfaces (sassi). Nat Lang Eng **6**(3&4)
International Telecommunication Union (ITU) (2003) Subjective quality evaluation of telephone services based on spoken dialogue systems
Jacko JA (ed) (2012) The human-computer interaction handbook: fundamentals, evolving technologies, and emerging applications, 3rd edn. CRC Press, Boca Raton
Kellar M, Watters C, Shepherd M (2006) A goal-based classification of web information tasks. In: In 69th annual meeting of the American society for information Science and Technology
Lazar J, Feng JH, Hochheiser H (2010) Research methods in human-computer interaction. Wiley, New York
Lee C, Jung S, Kim S, Lee GG (2009) Example-based dialog modeling for practical multi-domain dialog system. Speech Commun 51(5):466–484
Nakano M, Sato S, Komatani K, Matsuyama K, Funakoshi K, Okuno HG (2011) A two-stage domain selection framework for extensible multi-domain spoken dialogue systems. In: Proceedings of SIGdial. Association of Computational Linguistics, Stroudsburg, PA
Pappu, A, Rudnicky AI (2013) Predicting tasks in goal-oriented spoken dialog systems using semantic knowledge bases. In: Proceedings of SIGdial, Metz
Reichel S, Ehrlich U, Berton A, Weber M (2014) In-car multi-domain spoken dialogs: A wizard of oz study. In: EACL workshop dialog in motion, Gothenburg
Reichel S, Sohn J, Ehrlich U, Berton A, Weber M (2014) Out-of-domain spoken dialogs in the car: A woz study. In: Proceedings of SIGdial, Philadelphia, PA
Reithinger N, Alexandersson J, Becker T, Blocher A, Engel R, Löckelt M, Müller J, Pfleger N, Poller P, Streit M, Tschernomas V (2003) Smartkom: adaptive and flexible multimodal access to multiple applications. In: Multimodal interfaces, New York
Robichaud JP, Crook PA, Xu P, Khan OZ, Sarikaya R (2014) Hypotheses ranking for robust domain classification and tracking in dialogue systems. In: Proceedings of INTERSPEECH
Skantze G (2014) Error handling in spoken dialogue systems. Ph.D. thesis, KTH Computer Science and Communication
Stoyanchev S, Liu A, Hirschberg J (2014) Towards natural clarification questions in dialogue systems. In: AISB symposium on questions, discourse and dialogue: 20 years after making it explicit, London
Suhm B, Myers B, Waibel A (2001) Multimodal error correction for speech user interfaces. ACM Trans Comput-Hum Interact 8(1):60–98
The Nielsen Company (2014) Smartphones: So many apps, so much time
Young S (2014) Keynote: statistical approaches to open-domain spoken dialogue systems. In: Proceedings of SIGdial, Philadelphia, PA
Zoltan-Ford E (1991) How to get people to say and type what computers can understand. Int J Man-Mach Stud 34:527–547

Chapter 4
Analysis of an Extended Interaction Quality Corpus

Stefan Ultes, María Jesús Platero Sánchez, Alexander Schmitt, and Wolfgang Minker

Abstract The interaction quality paradigm has been suggested as evaluation method for spoken dialogue systems and several experiments based on the LEGO corpus have shown its suitability. However, the corpus size was rather limited resulting in insufficient data for some mathematical models. Hence, we present an extension to the LEGO corpus. We validate the annotation process and further show that applying support vector machine estimation results in similar performance on the original, the new and the combined data. Finally, we test previous statements about applying a Conditioned Hidden Markov Model or Rule Induction classification using the new data set.

Keywords Automatic dialoge systems evaluation • Statistical classification • Support vector machine • Hidden markov model

4.1 Introduction

Assessing the performance of spoken dialogue systems (SDSs) is still an open issue, although research has been conducted in this field for over a decade. The task may be solved using objective and subjective criteria. Here, objective criteria contain measures like dialogue length or success rate which are easily measurable and offer a direct connection to commercial interests. Subjective criteria usually contain the user experience or the user satisfaction. While the latter two are unarguably in the focus of the system users, both are much harder to measure automatically.

Interaction quality (IQ) as defined by Schmitt et al. (2011) is another subjective criterion and may be regarded as a more objective version of user satisfaction. The

---

S. Ultes (✉) • A. Schmitt • W. Minker
Ulm University, Ulm, Germany
e-mail: stefan.ultes@uni-ulm.de; alexander.schmitt@uni-ulm.de; wolfgang.minker@uni-ulm.de

M.J. Platero Sánchez
University of Granada, Granada, Spain
e-mail: plasez@correo.ugr.es

G. Geunbae Lee et al. (eds.), *Natural Language Dialog Systems and Intelligent Assistants*, DOI 10.1007/978-3-319-19291-8_4

main difference is that instead of asking the actual users, experts rate the dialogues. In previous work, we have shown that interaction quality may well be used instead of user satisfaction (Ultes et al. 2013b). A number of automatic estimation approaches have been investigated by us (Schmitt et al. 2011; Ultes et al. 2012a; Ultes and Minker 2013a, 2014) and others (El Asri et al. 2014). Our focus, however, was on applying IQ for online-adaption of the dialogue (Ultes et al. 2011, 2012b, 2014a,b).

However, the size of the available data in the *LEGO* corpus (Schmitt et al. 2012) for the experiments posed a critical limitation especially for experiments casting the problem as a sequential classification task (Ultes et al. 2012a). Hence, in this contribution, we present *LEGOext*, an extension of the *LEGO* corpus.[1] We compare the corpus characteristics of both the original and the new data in order to validate the labeling process. We analyze the performance of previously applied classification approaches on the new extended feature set. Furthermore, we compare the classification performance on the old and new data including cross-corpus analysis.

The outline of this work is as follows: the general idea of the interaction quality paradigm is presented in Sect. 4.2 including a brief description of the original *LEGO* corpus. The extension of this corpus along with an extended analysis and validation of the annotation process is presented in Sect. 4.3. Several different classification methods are applied and evaluated in Sect. 4.4 followed by a short discussion of the findings in Sect. 4.5.

4.2 The Interaction Quality Paradigm

The general idea of the interaction quality (IQ) paradigm—IQ being defined as user satisfaction annotated by expert raters—is to derive a number of interaction parameters from the dialogue system and use those as input variables to train a statistical classifier targeting IQ. Interaction quality is modeled on a scale from 5 to 1 representing the ratings "satisfied" (5), "slightly unsatisfied" (4), "unsatisfied" (3), "strongly unsatisfied" (2), and "extremely unsatisfied" (1).

The IQ paradigm originally presented by Schmitt et al. (2011) is based on automatically deriving interaction parameters from the SDS and feed these parameters into a statistical classification module which predicts the IQ level of the ongoing interaction at the current system-user exchange. The interaction parameters are rendered on three levels (see Fig. 4.1): the exchange level, the window level, and the dialogue level. The exchange level comprises parameters derived from SDS modules Automatic Speech Recognition (ASR), Spoken Language Understanding (SLU), and Dialogue Management (DM) directly. Parameters on the window and the dialogue level are sums, means, frequencies, or counts of exchange level parameters. While dialogue level parameters are computed out of all exchanges of the dialogue up to the current exchange, window level parameters are only computed out of the last three exchanges.

---

[1] *LEGOext* and *LEGO* are publicly available under http://nt.uni-ulm.de/ds-lego.

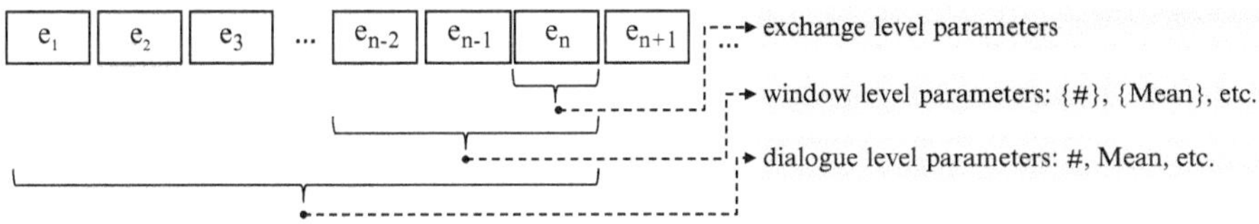

Fig. 4.1 This figure originally published by Schmitt et al. (2011) shows the three parameter levels constituting the interaction parameters: the exchange level containing information about the current exchange, the window level containing information about the last three exchanges, and the dialogue level containing information about the complete dialogue up to the current exchange

These interaction parameters are used as input variables to a statistical classification module. The statistical model is trained based on annotated dialogues of the Lets Go Bus Information System in Pittsburgh, USA (Raux et al. 2006). For the original *LEGO* corpus (Schmitt et al. 2012), 200 calls from 2006 consisting of 4885 exchanges have been annotated by three different raters resulting in a rating agreement of $\kappa = 0.54$.[2] Furthermore, the raters had to follow labeling guidelines to enable a consistent labeling process (Schmitt et al. 2012).

4.3 Corpus Statistics

In order to extend the *LEGO* corpus, an additional 201 calls to the Let's Go Bus Information System from 2007 consisting of 4753 exchanges have been annotated to constitute the *LEGOext* corpus. Three different raters being advanced students of computer science were asked to annotate each system-user exchange with one out of five satisfaction labels. They used an online form (Fig. 4.2) showing the complete call providing system output and user input as well as audio recordings of each user utterance and of the complete calls. Following the same rating guidelines as in the original *LEGO* corpus (Schmitt et al. 2012), the three raters achieved an overall inter-rater agreement of $\kappa = 0.5$.[2] General statistics for both corpora as well as the combined corpus *LEGOv2* are depicted in Table 4.1.

Comparing the agreement κ and correlation ρ[3] of the individual IQ ratings between the two corpora depicted in Table 4.2 shows that the annotation process using the guidelines results in similar agreement.

Since the aim is to model a general opinion on interaction quality, i.e., mirroring the IQ score other raters (and eventually users) agree with, the final label is determined empirically. Majority voting for deriving the final IQ label is not

[2]UAR, κ, and ρ are defined in Sect. 4.4.1

[3]UAR, κ, and ρ are defined in Sect. 4.4.1.

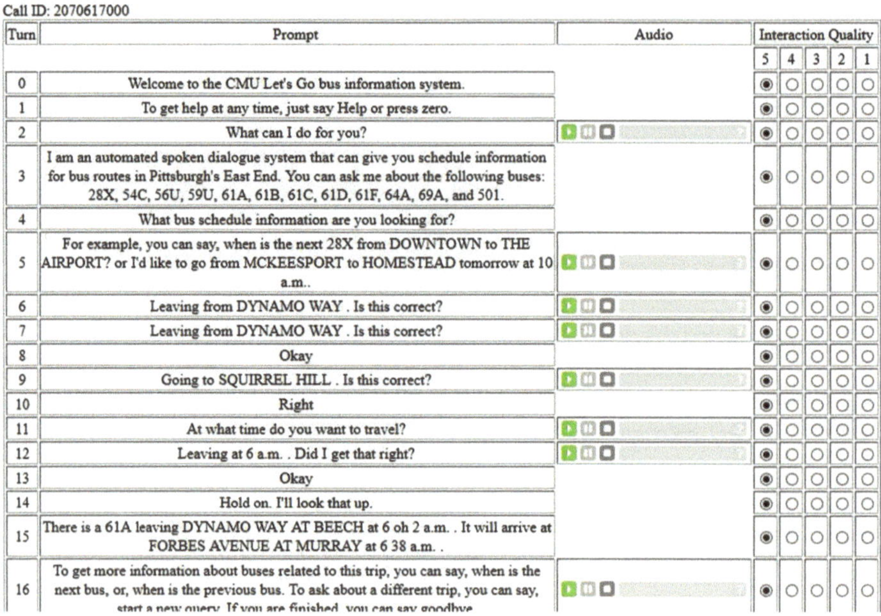
Call ID: 2070617000

Turn	Prompt	Audio	Interaction Quality
			5 4 3 2 1
0	Welcome to the CMU Let's Go bus information system.		
1	To get help at any time, just say Help or press zero.		
2	What can I do for you?		
3	I am an automated spoken dialogue system that can give you schedule information for bus routes in Pittsburgh's East End. You can ask me about the following buses: 28X, 54C, 56U, 59U, 61A, 61B, 61C, 61D, 61F, 64A, 69A, and 501.		
4	What bus schedule information are you looking for?		
5	For example, you can say, when is the next 28X from DOWNTOWN to THE AIRPORT? or I'd like to go from MCKEESPORT to HOMESTEAD tomorrow at 10 a.m..		
6	Leaving from DYNAMO WAY . Is this correct?		
7	Leaving from DYNAMO WAY . Is this correct?		
8	Okay		
9	Going to SQUIRREL HILL . Is this correct?		
10	Right		
11	At what time do you want to travel?		
12	Leaving at 6 a.m. . Did I get that right?		
13	Okay		
14	Hold on. I'll look that up.		
15	There is a 61A leaving DYNAMO WAY AT BEECH at 6 oh 2 a.m. . It will arrive at FORBES AVENUE AT MURRAY at 6 38 a.m. .		
16	To get more information about buses related to this trip, you can say, when is the next bus, or, when is the previous bus. To ask about a different trip, you can say, start a new query. If you are finished, you can say goodbye.		

Fig. 4.2 The online form used by the expert raters for annotating the *LEGOext* corpus

Table 4.1 Statistics of the two corpora *LEGO* and *LEGOext* and of the combined corpus *LEGOv2*

Corpus	Year	#calls	#exchanges	Avg. length	κ
LEGO	2006	200	4885	25.4	0.54
LEGOext	2007	201	4753	22.6	0.50
LEGOv2		401	9638	24.0	0.52

Shown are the recording year, the number of calls, the number of exchanges, the average dialogue length in number of exchanges, and the inter-rater agreement

Table 4.2 Agreement (κ) and correlation (ρ) in IQ ratings of the three raters in *LEGO* and *LEGOext*

LEGOext					*LEGO*				
	R1/R2	R1/R3	R2/R3	Mean		R1/R2	R1/R3	R2/R3	Mean
κ	0.40	0.51	0.59	**0.50**	κ	0.64	0.48	0.51	**0.54**
ρ	0.67	0.66	0.73	**0.69**	ρ	0.79	0.68	0.70	**0.72**

Expert ratings show similar correlations among each other

applicable since many exchanges are labeled with three different ratings, i.e., each of the three raters opted for a different score, thus forming no majority for either score. Therefore, the mean of all rater opinions is considered as possible candidate for the final class label:

Table 4.3 Agreement of single rater opinions to the merged label when determined by mean and median, measured in UAR, κ, and ρ

LEGOext			*LEGO*		
	Mean label	Median label		Mean label	Median label
UAR			*UAR*		
Rater1	0.550	0.648	Rater1	0.623	0.737
Rater2	0.410	0.512	Rater2	0.612	0.720
Rater3	0.600	0.844	Rater3	0.545	0.605
Mean	0.520	**0.668**	Mean	0.593	**0.687**
Cohen's weighted κ			*Cohen's weighted* κ		
Rater1	0.612	0.806	Rater1	0.763	0.815
Rater2	0.507	0.577	Rater2	0.767	0.814
Rater3	0.493	0.601	Rater3	0.657	0.658
Mean	0.539	**0.661**	Mean	0.729	**0.762**
Spearman's ρ			*Spearman's* ρ		
Rater1	0.843	0.891	Rater1	0.901	0.900
Rater2	0.905	0.846	Rater2	0.911	0.907
Rater3	0.782	0.799	Rater3	0.841	0.814
Mean	0.843	**0.845**	Mean	**0.884**	0.874

On the left side is *LEGOext*, on the right side *LEGO*

$$rating_{mean} = \lfloor \left(\frac{1}{R} \sum_{r=1}^{R} IQ_r \right) + 0.5 \rfloor \ . \tag{4.1}$$

Here, IQ_r is the interaction quality score provided by rater r. $\lfloor y \rfloor$ denotes the highest integer value smaller than y. Every value IQ_r contributes equally to the result that is finally rounded to the closest integer value.

Furthermore, the median is considered, which is defined as

$$rating_{median} = select(sort(IQ_r), \frac{R+1}{2}) \ , \tag{4.2}$$

where *sort* is a function that orders the ratings IQ_r of all R raters ascendingly and $select(list, i)$ chooses the item with index i from the list *list*. In other words, the IQ score separating the higher half of all ratings to the lower half is selected as final IQ score.

Table 4.3 shows the agreement between the mean and median labels with the single user ratings. Clearly, the median represents the better choice of final label given the higher values in κ, ρ, and unweighted average recall (UAR).[4] This validates the findings for the original experiments in the *LEGO* corpus.

[4]UAR, κ, and ρ are defined in Sect. 4.4.1.

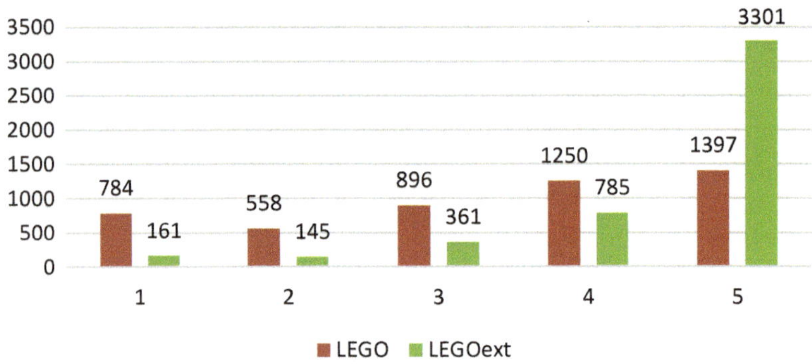

Fig. 4.3 The distribution of the final label scores along with the absolute number of occurrences for the *LEGO* and the *LEGOext* corpus

The distribution of the final IQ label is shown in Fig. 4.3. For the *LEGOext* corpus, label "5" has been assigned much more frequently while all others have been assigned less often compared to the *LEGO* corpus. This increase in overall system performance may be a result of an improved system as the 2007 version of Let's Go represents an updated system.

Naturally, this also results in a higher average IQ score for the *LEGOext* corpus: it achieves an average IQ of 4.46 while the *LEGO* corpus achieves 3.39 averaged over all labelled system-user exchanges.

4.4 IQ Modelling

For evaluating the performance of IQ with the new data set, three classification algorithms have been applied. The main evaluation has been conducted using a Support Vector Machine (SVM) (Vapnik 1995) with linear Kernel in accordance to Schmitt et al. (2011). Furthermore, IQ recognition has been cast as a sequence recognition problem with a Conditioned Hidden Markov Model (CHMM) (Ultes et al. 2011) using the JaCHMM library (Ultes et al. 2013a). A difference between a CHMM and an HMM is that a CHMM directly predicts a class probability $p(\omega|\mathbf{x}, \lambda)$ for sequence $\mathbf{x}$ while a conventional HMM only provides a probability $p(\mathbf{x}|\lambda)$ that the given model λ represents the observation sequence $\mathbf{x}$. The CHMM was included as initial tests have resulted in bad performance which was attributed to having not enough data (Ultes et al. 2012a). Finally, experiments using Rule Induction (RI) (Cohen 1995) are conducted.

The SVM experiments were conducted using tenfold cross-validation on the exchange level, i.e., the exchanges were assigned to one of ten subsets without regarding the call they belong to. In each fold, one subset is selected for evaluation while the remaining nine are used for training. By that, each sample is used for evaluation without having it within the training set at the same time. As the CHMM is based on the IQ value evolving over the course of the dialogue, sixfold cross-validation on the call-level has been applied. Here, each complete call has been assigned to one out of six subsets.

4.4.1 Evaluation Metrics

Three commonly applied evaluation metrics will be used in this contribution: UAR, Spearman's Rho, and Cohen's Kappa. The latter two also represent a measure for similarity of paired data. All measures will be briefly described in the following:

Unweighted Average Recall The Unweighted Average Recall (UAR) is defined as the sum of all class-wise recalls r_c divided by the number of classes $|C|$:

$$UAR = \frac{1}{|C|} \sum_{c \in C} r_c \,. \tag{4.3}$$

Recall r_c for class c is defined as

$$r_c = \frac{1}{|R_c|} \sum_{i=1}^{|R_c|} \delta_{h_i r_i} \,, \tag{4.4}$$

where δ is the Kronecker-delta, h_i and r_i represent the corresponding hypothesis-reference-pair of rating i, and $|R_c|$ the total number of all ratings of class c. In other words, UAR for multi-class classification problems is the accuracy corrected by the effects of unbalanced data.

Cohen's Kappa To measure the relative agreement between two corresponding sets of ratings, the number of label agreements corrected by the chance level of agreement divided by the maximum proportion of times the labelers could agree is computed. κ is defined as

$$\kappa = \frac{p_0 - p_c}{1 - p_c} \,, \tag{4.5}$$

where p_0 is the rate of agreement and p_c is the chance agreement (Cohen 1960). As US and IQ are on an ordinal scale, a weighting factor w is introduced reducing the discount of disagreements the smaller the difference is between two ratings (Cohen 1968):

$$w = \frac{|r_1 - r_2|}{|r_{\max} - r_{\min}|} \,. \tag{4.6}$$

Here, r_1 and r_2 denote the rating pair and $r_{\max}$ and $r_{\min}$ the maximal and minimal rating. This results in $w = 0$ for agreement and $w = 1$ if the ratings have maximal difference.

Spearman's Rho The correlation of two variables describes the degree by that one variable can be expressed by the other. *Spearman's Rank Correlation Coefficient* is a nonparametric method assuming a monotonic function between the two variables (Spearman 1904). It is defined by

$$\rho = \frac{\sum_i (x_i - \bar{x})(y_i - \bar{y})}{\sqrt{\sum_i (x_i - \bar{x})^2 \sum_i (y_i - \bar{y})^2}}, \tag{4.7}$$

where x_i and y_i are corresponding ranked ratings and $\bar{x}$ and $\bar{y}$ the mean ranks. Thus, two sets of ratings can have total correlation even if they never agree. This would happen if all ratings are shifted by the same value, for example.

4.4.2 Support Vector Machine

Three different experiments using a SVM have been conducted with the new data. First, the *LEGOext* corpus has been analyzed using different feature groups to identify their contribution to the overall performance. The AUTO group contains all (automatically derivable) features and subsumes the ASR, SLU, and DM feature groups which contain features belonging to the corresponding dialogue system module (cf. Sect. 4.2). The features used correspond to the list of features and their categorization of the *LEGO* corpus (Schmitt et al. 2012) and will not be restated here.

The results of SVM experiments on the *LEGOext* corpus are presented in Table 4.4 and show a UAR of 0.46 for the AUTO feature group. Furthermore, the results are compared with the performance of the *LEGO* corpus. It can be seen that, although *LEGOext* achieved lower performance, both corpora result in similar performances. Moreover, the DM feature group contributes most to the over all performance having ASR second and SLU third. This is notable as it shows that besides the ASR parameters, the DM parameters also have a major impact on the system performance.

A second experiment has been conducted using the combined *LEGOv2* corpus. The results are depicted in Table 4.5. With an overall performance of UAR 0.51 for the AUTO feature group, evaluating on the combined data achieves similar performance compared to each corpus separately. Evaluating the different feature groups furthermore also shows similar results compared to the performance on

Table 4.4 Results of SVM classification for all feature groups for each corpus separately

		LEGOext			*LEGO*		
	# feat.	UAR	κ	ρ	UAR	κ	ρ
ASR	29	0.378	0.287	0.494	0.458	0.535	0.689
SLU	5	0.221	0.093	0.239	0.260	0.219	0.311
DM	17	0.424	0.382	0.521	0.477	0.563	0.726
AUTO	51	0.463	0.482	0.604	0.512	0.614	0.764

Table 4.5 Results of SVM classification on the combined data set *LEGOv2*

	# feat.	UAR	κ	ρ
ASR	29	0.453	0.483	0.622
SLU	5	0.257	0.141	0.342
DM	17	0.446	0.443	0.538
AUTO	51	0.508	0.583	0.694

Table 4.6 Results of SVM classification trained on one corpus and evaluated on the other for all feature groups

	Train	Eval	UAR	κ	ρ
ASR			0.319	0.357	0.504
SLU	*LEGO*	*LEGOext*	0.275	0.239	0.372
DM			0.311	0.330	0.480
AUTO			0.331	0.379	0.554
ASR			0.302	0.129	0.441
SLU	*LEGOext*	*LEGO*	0.245	0.019	0.134
DM			0.441	0.257	0.474
AUTO			0.390	0.322	0.558

each corpus separately. However, for the combined data set, the *ASR* feature group contributes most to the overall performance.

Finally, the cross-corpus performance, i.e., training with one corpus and evaluating with the other corpus, has been investigated for all feature groups. Hence, no cross-validation has been applied. The results are depicted in Table 4.6. While performance decreases, the results are clearly above the majority baseline[5] for all feature groups. The finding that the DM parameters contribute most to the overall system performance is further emphasized: using only those parameters yield the best cross-corpus performance. This means that these feature groups contribute most to the generalization ability of the IQ paradigm.

4.4.3 Conditioned Hidden Markov Model

As previous studies investigating the applicability of the CHMM for IQ recognition resulted in low performance presumably due to lack of data, the *LEGOv2* corpus has been used to repeat the original experiments of Ultes et al. (2012a). The results are shown in Table 4.7 along with the results of the original experiment. Unfortunately, the performance has not increased. Two possible reasons have been identified: either the amount of data is still not sufficient or the CHMM is not a suitable model for IQ estimation. The latter might be attributed to the choice of Gaussian mixture models to model the observation probability.

[5]Majority baseline means that the majority class is always predicted. This would result in a UAR of 0.2 for a five class problem.

Table 4.7 Results of CHMM classification using the *LEGOv2* corpus compared with previous results of the *LEGO* corpus only (Ultes et al. 2012a)

	LEGOv2			*LEGO*		
# HS	UAR	κ	ρ	UAR	κ	ρ
5	0.39	0.399	0.542	0.38	0.4	0.56
6	0.379	0.405	0.562	0.38	0.39	0.57
7	0.376	0.402	0.561	0.35	0.4	0.59
8	0.336	0.27	0.385	0.37	0.41	0.59
9	0.394	0.406	0.562	0.39	0.43	0.6
10	0.38	0.412	0.567	0.37	0.39	0.55
11	0.389	0.417	0.566	0.36	0.41	0.58

Table 4.8 Performance of Rule Induction for cross-corpus evaluation

Train	Eval	UAR	κ	ρ
LEGOext	*LEGO*	0.374	0.235	0.513
LEGO	*LEGOext*	0.293	0.264	0.436

4.4.4 Rule Induction

As Rule Induction has shown to perform better than SVMs in previous work (Ultes and Minker 2014), RI has also been applied for IQ recognition. However, the claim was that RI produces a lot of specialized rules which result in worse generalizability of the model (Ultes and Minker 2013b). To investigate this, the cross-corpus experiment has been repeated using RI as a classification method. Again, no cross-validation has been applied due to the experiment characteristics. The results in Table 4.8 clearly show that RI achieves lower performance on the cross-corpora task for the `AUTO` feature set compared to the SVM. This confirms that using RI results in specialized models not as capable of generalizing than the SVM.

4.5 Discussion and Conclusion

In this work, we have presented an extension to the *LEGO* corpus adding 201 calls taken from the Let's Go Bus Information System in Pittsburgh, PA, USA. The new calls have been annotated with IQ labels from three different expert raters. The annotation statistics were similar to the statistics of the original corpus thus validating the annotation procedure. This has been underpinned by the performance of SVM classification of IQ on different feature groups achieving a UAR of 0.5 on the combined feature set. Furthermore, cross-corpus classification experiments have been conducted showing the transferability of IQ recognition for different system versions. The DM feature group has been identified as having a major contribution to IQ recognition performance both for evaluation within the corpus as well as for cross-corpus evaluation. Finally, a CHMM has shown to not increase performance having more data and Rule Induction has shown to be not as generalizable as SVMs thus validating claims in previous work.

References

Cohen J (1960) A coefficient of agreement for nominal scales. In: Educational and psychological measurement, vol 20, pp 37–46
Cohen J (1968) Weighted kappa: nominal scale agreement provision for scaled disagreement or partial credit. Psychol Bull 70(4):213
Cohen WW (1995) Fast effective rule induction. In: Proceedings of the 12th international conference on machine learning. Morgan Kaufmann, San Francisco, pp 115–123
El Asri L, Khouzaimi H, Laroche R, Pietquin O (2014) Ordinal regression for interaction quality prediction. In: IEEE international conference on acoustics, speech and signal processing (ICASSP). IEEE, Florence, pp 3245–3249
Raux A, Bohus D, Langner B, Black AW, Eskenazi M (2006) Doing research on a deployed spoken dialogue system: one year of let's go! experience. In: Proc. of the international conference on speech and language processing (ICSLP)
Schmitt A, Schatz B, Minker W (2011) Modeling and predicting quality in spoken human-computer interaction. In: Proceedings of the SIGDIAL 2011 conference. Association for Computational Linguistics, Portland, pp 173–184
Schmitt A, Ultes S, Minker W (2012) A parameterized and annotated spoken dialog corpus of the cmu let's go bus information system. In: International conference on language resources and evaluation (LREC), pp 3369–337
Spearman CE (1904) The proof and measurement of association between two things. Am J Psychol 15:88–103
Ultes S, Minker W (2013a) Improving interaction quality recognition using error correction. In: Proceedings of the 14th annual meeting of the special interest group on discourse and dialogue. Association for Computational Linguistics, Metz, pp 122–126. http://www.aclweb.org/anthology/W/W13/W13-4018
Ultes S, Minker W (2013b) Interaction quality: a review. SibSAU (as in Siberian State Aerospace University) Newspaper 4:153–156. http://www.vestnik.sibsau.ru/images/vestnik/ves450.pdf
Ultes S, Minker W (2014) Interaction quality estimation in spoken dialogue systems using hybrid-hmms. In: Proceedings of the 15th annual meeting of the special interest group on discourse and dialogue (SIGDIAL). Association for Computational Linguistics, Philadelphia, pp 208–217. http://www.aclweb.org/anthology/W14-4328
Ultes S, Heinroth T, Schmitt A, Minker W (2011) A theoretical framework for a user-centered spoken dialog manager. In: Proceedings of the paralinguistic information and its integration in spoken dialogue systems workshop. Springer, New York, pp. 241–246
Ultes S, ElChabb R, Minker W (2012a) Application and evaluation of a conditioned hidden markov model for estimating interaction quality of spoken dialogue systems. In: Mariani J, Devillers L, Garnier-Rizet M, Rosset S (eds) Proceedings of the 4th international workshop on spoken language dialog system (IWSDS). Springer, New York, pp 141–150
Ultes S, Schmitt A, Minker W (2012b) Towards quality-adaptive spoken dialogue management. In: NAACL-HLT workshop on future directions and needs in the spoken dialog community: tools and data (SDCTD 2012). Association for Computational Linguistics, Montréal, pp 49–52. http://www.aclweb.org/anthology/W12-1819
Ultes S, ElChabb R, Schmitt A, Minker W (2013a) Jachmm: a java-based conditioned hidden markov model library. In: IEEE international conference on acoustics, speech and signal processing (ICASSP). IEEE, Vancouver, pp 3213–3217
Ultes S, Schmitt A, Minker W (2013b) On quality ratings for spoken dialogue systems – experts vs. users. In: Proceedings of the 2013 conference of the North American chapter of the association for computational linguistics: human language technologies. Association for Computational Linguistics, Atlanta, pp 569–578

Ultes S, Dikme H, Minker W (2014a) Dialogue management for user-centered adaptive dialogue. In: Proceedings of the 5th international workshop on spoken dialogue systems (IWSDS)
Ultes S, Dikme H, Minker W (2014b) First insight into quality-adaptive dialogue. In: International conference on language resources and evaluation (LREC), pp 246–251
Vapnik VN (1995) The nature of statistical learning theory. Springer, New York

Chapter 5
HALEF: An Open-Source Standard-Compliant Telephony-Based Modular Spoken Dialog System: A Review and An Outlook

David Suendermann-Oeft, Vikram Ramanarayanan, Moritz Teckenbrock, Felix Neutatz, and Dennis Schmidt

Abstract We describe completed and ongoing research on HALEF, a telephony-based open-source spoken dialog system that can be used with different plug-and-play back-end modules. We present two examples of such a module, one which classifies whether the person calling into the system is intoxicated or not and the other a question answering application. The system is compliant with World Wide Web Consortium and related industry standards while maintaining an open codebase to encourage progressive development and a common standard testbed for spoken dialog system development and benchmarking. The system can be deployed towards a versatile range of potential applications, including intelligent tutoring, language learning and assessment.

Keywords Spoken dialog systems • VoiceXML • Alcoholic language classification

5.1 Introduction

Spoken dialog systems (SDSs) have witnessed a steep increase in usage over the last 5 years thanks to improvements in speech recognition performance, the availability of smart devices, ubiquitous high-speed internet and cloud computing, progress in developing standards and the emergence of crowdsourcing for speech applications, among other factors (Suendermann-Oeft 2014). While commercially deployed industrial vendors (such as Cisco, Nuance, Avaya, Genesys, Microsoft, Voxeo, etc.) tend to concentrate on dialog managers with finite-state call flows,

---

Part of the work described here was completed when the first author was at DHBW Stuttgart.

D. Suendermann-Oeft • V. Ramanarayanan (✉)
Educational Testing Service (ETS) Research, San Francisco, CA, USA
e-mail: suendermann-oeft@ets.org; vramanarayanan@ets.org

M. Teckenbrock • F. Neutatz • D. Schmidt
DHBW, Stuttgart, Germany

G. Geunbae Lee et al. (eds.), *Natural Language Dialog Systems and Intelligent Assistants*, DOI 10.1007/978-3-319-19291-8_5

rule-based grammars for speech recognition, large volumes of call data and more or less standardized interfaces and protocols, academic research (see, for example, Bos et al. 2003; Raux et al. 2005; Williams and Young 2007; Bohus et al. 2007; Young et al. 2010; Black et al. 2011) has adopted a more long-term approach, focusing on statistically trained dialog managers and spoken language understanding modules, smaller-sized datasets and proprietary interfaces (Pieraccini and Huerta 2005; Suendermann 2011). Academia has also been more open to publishing software and research results as compared to industry. Having said that, a large percentage of practical, deployed solutions are industry-based and are, as such, proprietary. Although there are standard protocols in place to develop SDS solutions which many industrial systems adhere to, their system implementations and software components are often different, which makes benchmarking of systems relative to each other a difficult task. An open-source implementation that is compliant with W3C standards would be a positive step towards a working solution to this issue.

It is further important to note the utility of having a telephony-based, modular SDS architecture. Although there exist many open-source SDS implementations in the academic world, most of these are not telephony-based. This means that most of these systems typically require installation of a software interface on a local workstation or computer. A telephony-based SDS setup would allow people to call into and access the SDS without any software installation overhead. Furthermore, by making such a system modular, we can individually optimize the different components of the system—telephony server, speech server, voice browser and web server. However, there are currently almost no systems that offer all of the above advantages (Suendermann 2011) to our knowledge [the CMU Olympus (Bohus et al. 2007) and the ALEX dialog frameworks (Jurčíček et al. 2014) are two systems that come close to being exceptions, but that they are standard-compliant is not clear].

To address these shortcomings, we developed HALEF—a telephony-based, modular, open-source, standard-compliant spoken dialog system. The primary objective of this paper is to describe the current state of the HALEF system and discuss how various back-end applications can be integrated within the SDS framework. HALEF is written primarily in Java and leverages a number of open-source tools in a distributed framework for scalability.

SDS frameworks can be deployed to suit a wide range of applications such as directory services (Gorin et al. 1997), technical troubleshooting (Schmitt et al. 2010), intelligent tutoring (Graesser et al. 2005) or computer-assisted language learning (Seneff et al. 2004; Xu and Seneff 2011). In this paper, we present a couple of such applications, including that of a plug-and-play module for alcoholic state classification and a question answering service. Note however that the purpose of this paper is *not* to further the state of the art in alcohol classification or question answering applications, but to demonstrate how a working classifier/application can be incorporated into the HALEF framework as an independent plug-and-play module.

The rest of the paper is organized as follows: Sect. 5.2 describes the basic architecture and components of the HALEF spoken dialog system. We then describe example applications of the HALEF SDS to alcoholic state classification and question answering in Sect. 5.3. Finally we conclude with a discussion of ongoing and future research into the system in Sect. 5.4.

5.2 HALEF System Description

The HALEF (Help Assistant–Language-Enabled and Free) framework leverages different open-source components to form an SDS framework that is modular and industry-standard-compliant: Asterisk, a SIP—(Session Initiation Protocol) and PSTN—(Public Switched Telephone Network) compatible telephony server (van Meggelen et al. 2009); JVoiceXML, an open-source voice browser that can process SIP traffic (Schnelle-Walka et al. 2013) via a voice browser interface called Zanzibar (Prylipko et al. 2011); Cairo, an MRCP (Media Resource Control Protocol) speech server, which allows the voice browser to initiate SIP or RTP (Real-time Transport Protocol) connections from/to the telephony server (Prylipko et al. 2011); the Sphinx automatic speech recognizer (Lamere et al. 2003); Festival (Taylor et al. 1998) and Mary (Schröder and Trouvain 2003)—text-to-speech synthesis engines; and an Apache Tomcat-based web server that can host dynamic VoiceXML (VXML) pages and serve media files such as grammars[1] and audio files to the voice browser. Figure 5.1 schematically depicts the main components of the HALEF system. Note that unlike a typical SDS, which consists of sequentially connected modules for speech recognition, language understanding, dialog management, language generation and speech synthesis, in HALEF some of these are grouped together forming independent blocks which are hosted on different virtual machines in a distributed architecture. For further details on the individual blocks as well as design choices, please refer to Mehrez et al. (2013). In this framework, one can serve different back-end applications as standalone web services on a separate server. Incorporating the appropriate start URL (Universal Resource Locator) of the web service in the VXML input code that the voice browser interprets will then allow the voice browser to trigger the web application at the appropriate point in the callflow. The web services in our case typically take as input any valid HTTP-based GET or POST request and output a VXML page that the voice browser can process next.

In order to understand how HALEF works in a better manner, let us consider an example. Figure 5.2 illustrates the step-by-step flow of operations that are executed in the case of a question answering (QA) back-end application. Once the Asterisk server receives a call, it sends a notification to the voice browser to fetch the VXML code from the web server. The voice browser in turn identifies the resources that the speech server will need to prepare for this application. It then notifies the MRCP server and starts sessions and channels for all required resources including the

[1]Popular grammar formats include JSGF (Java Speech Grammar Format), SRGS (speech recognition grammar specification) and ARPA (Advanced Research Projects Agency) formats.

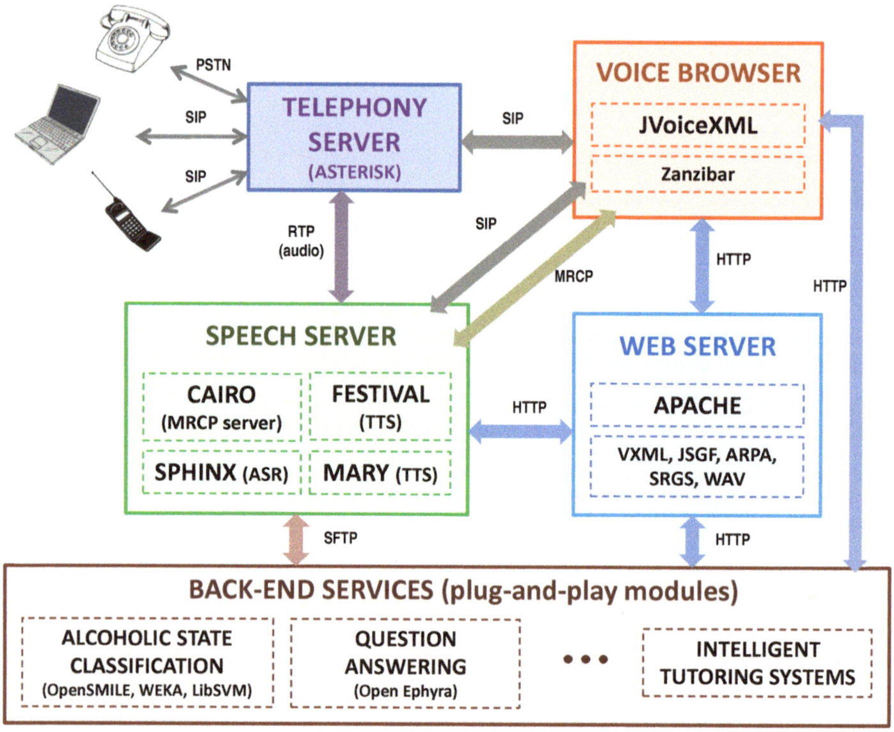

Fig. 5.1 System architecture of the HALEF spoken dialog system depicting the various modular open-source components

provisioning of speech recognition grammars. Finally, the speech server sends a SIP response back to the voice browser and Asterisk to confirm session initiation. Completion of this process successfully establishes a communication channel between the user and Halef's components.

Now that the session is established, Asterisk streams audio via RTP to the speech server. When the caller starts speaking, the Sphinx engine's voice activity detector fires and identifies voiced portions of the speech and starts decoding these portions. When the voice activity detector finds that the caller has finished speaking, Sphinx sends the recognition result back to the voice browser, which passes it on to the standalone QA web application (which is served on another server) via HTTP and waits for an answer. It then sends this answer to the dialog manager which evaluates and generates VXML code with the final response to be spoken out by the speech synthesizer (either Festival or Mary). The voice browser then interprets this VXML code and sends a synthesis request to the speech server with the response. Festival/Mary synthesizes the response and passes the result back via RTP to Asterisk, which forwards the audio signal to the user. At the same time, Cairo sends a confirmation signal to the voice browser. After receiving this signal, the voice browser sends a cleanup request to close all open channels and resources. This ends the SIP session with Asterisk, which finally triggers Asterisk to send an end-of-call signal to the user.

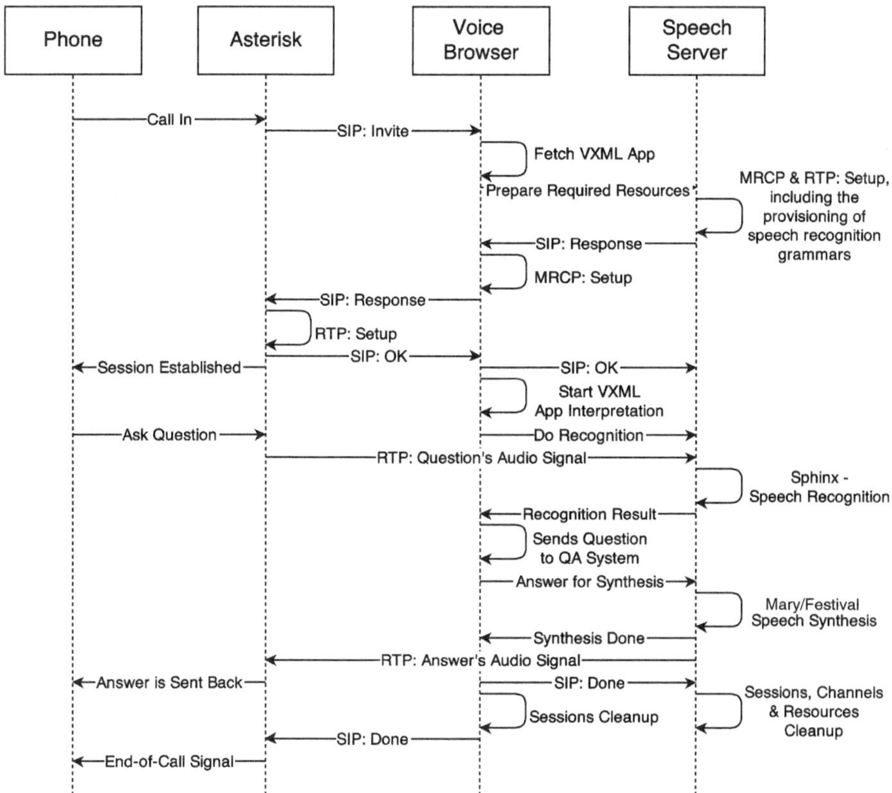

Fig. 5.2 Flow diagram of an example HALEF call flow for the question answering application

Note that HALEF makes no assumptions on the specifics of the dialog management system used. One could choose to use a specific rule-based call flow management routine (in which case one would have to generate VXML pages corresponding to actions for each rule branch of the routine) or a more statistical system, such as one based on Partially Observable Markov Decision Processes (which one could implement as a separate web service that returns an appropriate VXML page detailing the next action to be taken by the SDS). There is similar flexibility in designing aspects of the spoken language understanding and language models for speech recognition (or grammars). In case of the latter, one could imagine wanting to use different grammars depending on the language or the domain in question. Currently HALEF supports the use of either JSGF (Java Speech Grammar Format) and ARPA (Advanced Research Projects Agency) formats to specify grammars. This modularity in design is intended to allow users more flexibility and ease of use in adapting HALEF to different use cases and environments.

5.3 Specific Back-End Use Case Examples

5.3.1 Case Study I: A Question Answering Application

The flow diagram in Fig. 5.2 depicts the sequence of operations executed in the case of a back-end interface that allows HALEF to interact with a question answering (QA) web application called OpenEphyra (van Zaanen 2008), which was developed by researchers working on the IBM Watson DeepQA initiative (Ferrucci et al. 2010). We shall only briefly mention the key features here—for further details please see Mehrez et al. (2013). The application is a combination of several components including question analysis, query generation, pattern matching, answer extraction and answer selection. As this system has already been elucidated in the publications cited above, we only provide a brief description of the steps involved in answering a question here. First, the spoken input question is normalized for punctuation, abbreviations, etc. and then stemmed for nouns and verbs. Next, keywords, question type and named entities are extracted to form queries that are subsequently used to search the available knowledge base. After matching possible candidates in the database, an n-best list is returned, following which the answer with the highest confidence is chosen as the output.

5.3.2 Case Study II: Alcoholic State Classification

In this section we present an example of a plug-and-play alcoholic state classification module that can be used with HALEF. The problem of alcoholic state classification has recently gained popularity in the pattern recognition community, leading to the proposal of competitions at academic conferences such as the Interspeech 2011 Speaker State Challenge (Schuller et al. 2011). That being said, recall that the main goal of the paper is *not* to present state-of-the-art classification results, but to present a working classification module (which can be optimized for performance independent of the HALEF system).

Similar to the previously described case of the question answering application, we served the alcohol language classifier as a *standalone* web service—or more specifically a Java servlet that is served by Apache Tomcat. The speech server ships the incoming audio file to this web service, which then performs three operations. First, it preprocesses the incoming audio file and extracts features using OpenSMILE. Then it uses Weka to perform the classification using a previously trained model. Finally, it extracts the result and generates a corresponding VXML page that contains information to be processed by the voice browser regarding how it should proceed further.

Table 5.1 List of speech prompts used from the Alcohol Language Corpus (Schiel et al. 2008; Schiel and Heinrich 2009) and their corresponding test classification performance (represented as unweighted average recall, UAR)

Exp.	Command	# samples	Test UAR (%)
1	Sportplatzweg 27, Marktgraitz	228	68
2	Temperature 23 °C	268	78
3	Nächster Titel	268	73
4	Frequency 92.2 MHz	268	60
5	Autobahn meiden	268	63

5.3.2.1 Data

We used the Alcohol Language Corpus (ALC) collected at the Ludwig Maximilians University of Munich to train the classifier. The dataset contains audio recordings of people in sober and alcohol-intoxicated state (Schiel et al. 2008; Schiel and Heinrich 2009), comprising 39 h of speech from 77 female and 85 male speakers. Out of this, we performed experiments on a reduced dataset[2] that was introduced by the Interspeech 2011 Speaker State challenge (Schuller et al. 2011). We further converted all audio instances of the ALC from 44.1 to 8 kHz sample rate to ensure compatibility with HALEF.

Classification tasks that leverage speech collected using a spoken dialog system are bound to certain constraints. For example, speaker turns cannot be arbitrarily long in duration in a practical setting. This is even more so when one is testing for alcohol intoxication. Therefore we only considered experimental trials during which speakers spoke prompts that were short in duration. We chose five speech prompts from the ALC that met these requirements—see Table 5.1 for a list of these prompts.

5.3.2.2 Classification Paradigm

In order to classify as sober or alcohol intoxicated, the test person dials into the HALEF system and is prompted to repeat one of the prompts, for example "*Temperature 23 °C*" . As mentioned earlier, after the user input has been recorded, a web service is triggered to run openSMILE (Eyben et al. 2010) to generate a sequence of feature vectors. The acoustic feature set used corresponds to the configuration of the Interspeech 2011 Speaker State Challenge—4368 features comprising a multitude of low-level descriptors (such as spectral features, F0, etc.) and their applied functionals; see Schuller et al. (2011) for more details.

[2]Since the data collected during different ALC experiments are not balanced in terms of class and gender, we removed all speakers that were recorded in only one of the classification states. We then discarded as many male speakers (selected at random) as necessary to achieve gender balance.

We used support vector machine (SVM) classifiers to perform the classification. We ran all experiments with the Weka machine learning toolkit (version 3.7) (Holmes et al. 1994; Hall et al. 2009) in combination with LibSVM, an open-source implementation of support vector machines (Chang and Lin 2011). For evaluation we selected ten male/female speaker pairs as test set. We tuned the complexity parameter of the linear kernel by using leave-one-*speaker pair*-out cross-validation on the remaining speaker pairs. Table 5.1 lists the unweighted average recall (UAR) for each test prompt. We observe that although the system performs consistently better than chance, there is scope for improvement. However, we deemed it to be sufficient in order to set up a working prototype spoken dialog interface for our purposes.

5.4 Conclusions and Outlook

We have presented the state of the art of the HALEF system—a fully open-source, modular, telephony-based industry-standard-compliant spoken dialog system that can be interfaced with a number of potential back-end applications. We illustrated this capability with two example applications, one of alcoholic state classification and the other of a question answering application. HALEF can be accessed online at the following URL: http://halef.org. One can also call into HALEF for a demo at the following US-based telephone number: `(206) 203-5276 (Ext. 2000: QA demo; 2001: ALC demo)`. Another back-end application that we are currently developing is a system for English language learning and assessment tailored to address the conversational competency of a user.

References

Black AW, Burger S, Conkie A, Hastie H, Keizer S, Lemon O, Merigaud N, Parent G, Schubiner G, Thomson B, Williams J, Yu K, Young S, Eskenazi M (2011) Spoken dialog challenge 2010: comparison of live and control test results. In: Proceedings of the SIGDIAL 2011 conference. Association for Computational Linguistics, Portland, pp 2–7

Bohus D, Raux A, Harris T, Eskenazi M, Rudnicky A (2007) Olympus: an open-source framework for conversational spoken language interface research. In: Proc. of the HLT-NAACL, Rochester, 2007

Bos J, Klein E, Lemon O, Oka T (2003) Dipper: description and formalisation of an information-state update dialogue system architecture. In: 4th SIGdial workshop on discourse and dialogue, pp. 115–124

Chang CC, Lin CJ (2011) LIBSVM: a library for support vector machines. ACM Trans Intell Syst Technol 2(3):27

Eyben F, Wöllmer M, Schuller B (2010) Opensmile: the Munich versatile and fast open-source audio feature extractor. In: Proc. of the MM, Florence, 2010

Ferrucci D, Brown E, Chu-Carroll J, Fan J, Gondek D, Kalyanpur A, Lally A, Murdock W, Nyberg E, Prager J, Schlaefer N, Welty C (2010) Building Watson: an overview of the DeepQA project. AI Mag 31(3):59–79

Gorin A, Riccardi G, Wright J (1997) How may I help you? Speech Commun 23(1/2):113–127
Graesser AC, Chipman P, Haynes BC, Olney A (2005) Autotutor: an intelligent tutoring system with mixed-initiative dialogue. IEEE Trans Educ 48(4):612–618
Hall M, Frank E, Holmes G, Pfahringer B, Reutemann P, Witten IH (2009) The weka data mining software: an update. ACM SIGKDD Explor Newsl 11(1):10–18
Holmes G, Donkin A, Witten IH (1994) Weka: a machine learning workbench. In: Proceedings of the 1994 second Australian and New Zealand conference on intelligent information systems. IEEE, Brisbane, pp 357–361
Jurčíček F, Dušek O, Plátek O, Žilka L (2014) Alex: a statistical dialogue systems framework. In: Text, speech and dialogue. Springer, Brno, pp 587–594
Lamere P, Kwok P, Gouvea E, Raj B, Singh R, Walker W, Warmuth M, Wolf P (2003) The CMU SPHINX-4 speech recognition system. In: Proc. of the ICASSP'03, Hong Kong, 2003
Mehrez T, Abdelkawy A, Heikal Y, Lange P, Nabil H, Suendermann-Oeft D (2013) Who discovered the electron neutrino? A telephony-based distributed open-source standard-compliant spoken dialog system for question answering. In: Proc. of the GSCL, Darmstadt, 2013
Pieraccini R, Huerta J (2005) Where do we go from here? Research and commercial spoken dialog systems. In: Proc. of the SIGdial, Lisbon, 2005
Prylipko D, Schnelle-Walka D, Lord S, Wendemuth A (2011) Zanzibar OpenIVR: an open-source framework for development of spoken dialog systems. In: Proc. of the TSD, Pilsen
Raux A, Langner B, Bohus D, Black A, Eskenazi M (2005) Let's go public! taking a spoken dialog system to the real world. In: Proc. of the Interspeech, Lisbon, 2005
Schiel F, Heinrich C (2009) Laying the foundation for in-car alcohol detection by speech. In: Proc. of the Interspeech, Brighton, 2009
Schiel F, Heinrich C, Barfüsser S, Gilg T (2008) ALC—alcohol language corpus. In: Proc. of the LREC, Marrakesh, 2008
Schmitt A, Scholz M, Minker W, Liscombe J, Suendermann D (2010) Is it possible to predict task completion in automated troubleshooters? In: Proc. of the Interspeech, Makuhari, 2010
Schnelle-Walka D, Radomski S, Mühlhäuser M (2013) JVoiceXML as a modality component in the W3C multimodal architecture. J Multimodal User Interfaces 7:183–194
Schröder M, Trouvain J (2003) The German text-to-speech synthesis system mary: a tool for research, development and teaching. Int J Speech Technol 6(4):365–377
Schuller B, Steidl S, Batliner A, Schiel F, Krajewski J (2011) The interspeech 2011 speaker state challenge. In: INTERSPEECH, pp 3201–3204
Seneff S, Wang C, Zhang J (2004) Spoken conversational interaction for language learning. In: InSTIL/ICALL symposium
Suendermann D (2011) Advances in commercial deployment of spoken dialog systems. Springer, New York
Suendermann-Oeft D (2014) Modern conversational agents. In: Technologien für digitale Innovationen. Springer, Wiesbaden, pp 63–84
Taylor P, Black A, Caley R (1998) The architecture of the festival speech synthesis system. In: Proc. of the ESCA workshop on speech synthesis, Jenolan Caves, 1998
van Meggelen J, Smith J, Madsen L (2009) Asterisk: the future of telephony. O'Reilly, Sebastopol
van Zaanen M (2008) Multi-lingual question answering using OpenEphyra. In: Working notes for the cross language evaluation forum (CLEF), pp 1–6
Williams JD, Young S (2007) Partially observable markov decision processes for spoken dialog systems. Comput Speech Lang 21(2):393–422
Xu Y, Seneff S (2011) A generic framework for building dialogue games for language learning: application in the flight domain. In: SLaTE, pp 73–76
Young S, Gašić M, Keizer S, Mairesse F, Schatzmann J, Thomson B, Yu K (2010) The hidden information state model: a practical framework for pomdp-based spoken dialogue management. Comput Speech Lang 24(2):150–174

Chapter 6
Micro-Counseling Dialog System Based on Semantic Content

Sangdo Han, Yonghee Kim, and Gary Geunbae Lee

Abstract This paper introduces a text dialog system that can provide counseling dialog based on the semantic content of user utterances. We extract emotion-, problem-, and reason-oriented semantic contents from user utterances to generate micro-counseling system responses. Our counseling strategy follows micro-counseling techniques to build a working relationship with a client and to discover the client's concerns and problems. Extracting semantic contents allows the system to generate appropriate counseling responses for various user utterances. Experiments show that our system works well as a virtual counselor.

Keywords Dialog system • Counseling dialog system • Micro-counseling technique • Semantic content • Back-off strategy

6.1 Introduction

People often talk with other people to share their situation and to relieve stress. However, other people are not always available, and we may not want to reveal all information because some of it may be too personal; a micro-counseling dialog system can solve these problems. In our previous work, the system could not understand various user utterances because it used only lexical information to analyze them (Han et al. 2013). In this work, we developed a system that analyzes semantic information to achieve understanding of user utterances and to effectively respond to them for counseling.

In this paper, we measure the effect of our new information extracting method, new counseling information, and chat-oriented back-off strategy. Our system can extract information from a wider variety of utterances and get higher scores for counseling satisfaction than the previous system.

S. Han (✉) • Y. Kim • G.G. Lee
Pohang University of Science and Technology, Pohang, Republic of Korea
e-mail: hansd@postech.ac.kr; ttti07@postech.ac.kr; gblee@postech.ac.kr

G. Geunbae Lee et al. (eds.), *Natural Language Dialog Systems and Intelligent Assistants*, DOI 10.1007/978-3-319-19291-8_6

Relevant related work is presented in Sect. 6.2. Micro-counseling techniques are summarized in Sect. 6.3. Corpus data are introduced in Sect. 6.4, and the micro-counseling dialog method is described in Sect. 6.5. The experiments and results are shown in Sect. 6.6 and conclusion is drawn in Sect. 6.7.

6.2 Related Work

Han et al. (2013) used a conditional random field algorithm to extract "who, what, when, where, why, how" (5W1H) information to counsel, but because the system only considers 5W1H information, some system utterances that consider time and place are not relevant in a counseling dialog. For example, the system could generate utterance like "Where did you mad?" In addition, because the method is based on only lexical information, it needs a large corpus to understand various user utterances. Furthermore, this method could not detect various user emotions because it was based on only keyword matching.

Meguro et al. (2013) introduced a listening-oriented dialog system based on a model trained by a partially observable Markov decision process using human–human dialog corpus. The system uses a listening-oriented dialog strategy to encourage users to speak, but the system utterances are limited because it selects responses from the corpus. It also cannot respond to utterances that are not in the specific domain.

Using extracted emotion-, problem-, and reason-oriented information by extracting general semantic contents (subject, predicate, and object), then using this information to guide selection of appropriate counseling responses. By redefining counseling information from 5W1H, the system focuses on the user's current situation and emotional state. The new method extracts this information by analyzing general semantic contents, so it can extract the information from various domain-independent utterances. However, not all utterances are relevant sources of semantic contents for counseling, and the counseling system should respond to all user utterances in order to encourage the users to continue talking; in this case the system should adopt a "back-off strategy" in which it uses a chat-oriented system to respond with a relevant sentence that has no counseling value, but which encourages the client to continue interacting. Most chat-oriented systems (e.g., ELIZA (Weizenbaum 1966), ALICE[1]) are based on the simple pattern matching technique, but several systems are based on a sentence similarity measure (Lee et al. 2009; Li et al. 2004); they select the most similar sentence to the user input among example sentence pairs and generate modified sentence as an output.

[1]ALICE: Artificial Intelligence Foundation Inc. http://www.alicebot.org.

6.3 Micro-Counseling Techniques

Micro-counseling techniques are basic counseling techniques that make clients feel that a counselor listens carefully and understands the clients (Evans et al. 2010). Micro-counseling includes four main techniques: attending, paraphrasing, reflecting feelings, and questioning.

Attending is a technique to react naturally to an utterance. Attending utterances could follow any kind of user utterances. This technique makes a client feel that the system focuses on him or her and encourages the client to continue talking to the system. Examples include “Please tell me more” and “Continue.”

Paraphrasing is a technique to make the user think the system is following what the user said. Unlike attending, paraphrasing utterance is dependent on a user utterance because the system should rephrase the client’s utterance. For example, when client says “I ate pizza,” the counselor could say “Oh, you ate pizza.”

Reflecting feelings organizes the user’s whole situation. This technique is similar to paraphrasing but whereas paraphrasing follows exactly what the client said in the previous turn, reflecting feelings follows all information that the user provides. For example “You don’t feel good because John deleted it to stop it,” or “Stopping it made you sad.”

Questioning is a technique to ask a user to provide more counseling information, e.g. “How do you feel about it?,” or “Why did John do so?.”

6.4 Data Collection

We generated 512 utterances as a counseling corpus (Table 6.1). Because micro-counseling dialog is based on problems, feelings, and specific facts (Evans et al. 2010), our generated utterances focus on user’s problem, emotion and reason based on micro-counseling techniques. We generated the corpus based on 42 counseling situation (Table 6.2) and micro-counseling techniques. This corpus used to select micro-counseling utterances.

Table 6.1 Counseling corpus example

Speaker	Utterance
System	Hello. How are you today?
User	I feel bad because I fought with my boyfriend.
System	You fought with your boyfriend. Why does it happen?
User	He didn’t remember my birthday.
System	I see. You feel bad because he didn’t remember your birthday.
User	That’s right

Table 6.2 Example of counseling situations

Emotion	Problem	Reason
Angry	I fought with John.	John yelled at me.
Sad	My dog died.	He fell from cliff.
Happy	My dad won the prize.	He got the best score.

A general chatting corpus was generated based on seven domain-independent dialog acts; it includes 11,328 user utterances. The corpus was generated by collecting chatting dialog between two people. It was used for micro-counseling utterance detection.

To generate counseling information extraction rules, we used Movie-Dic, which is a movie script corpus from 753 movies (Banchs 2012). It includes 132,229 utterances, which we assume represent natural dialogs.

6.5 Method

6.5.1 Architecture

Our system consists of four components: counseling utterance understanding (CUU), counseling strategy managing (CSM), counseling response generating (CRG), and a chat-oriented back-off dialog system. CUU understands what a user says, CSM decides what kind of strategy to use, and CRG decides how to generate counseling utterances. The chat-oriented dialog system is used to respond to general user utterances for which counseling utterances are difficult to generate (Fig. 6.1).

6.5.2 Counseling Utterance Understanding

In the CUU module, the system first decides whether a user utterance is appropriate for micro-counseling dialog, then extracts counseling information. If the user utterance is not appropriate for a micro-counseling reaction, the chat-oriented dialog system generates a general response as back-off strategy.

Our system treats the utterances whose dialog act is a statement as appropriate utterances for micro-counseling dialog. Semantic contents to generate counseling response are mostly included in utterances whose dialog act is a statement because their purposes are to deliver information. To detect a statement dialog act, we used the MaxEnt algorithm (Beger et al. 1996) using a chatting corpus which is labeled with dialog act. We trained a model with word and Part of Speech (POS) bi-gram features to train the model.

As a second step, we check whether or not our system can extract semantic contents from the user utterance. If it cannot, the utterance is passed to the chat-oriented

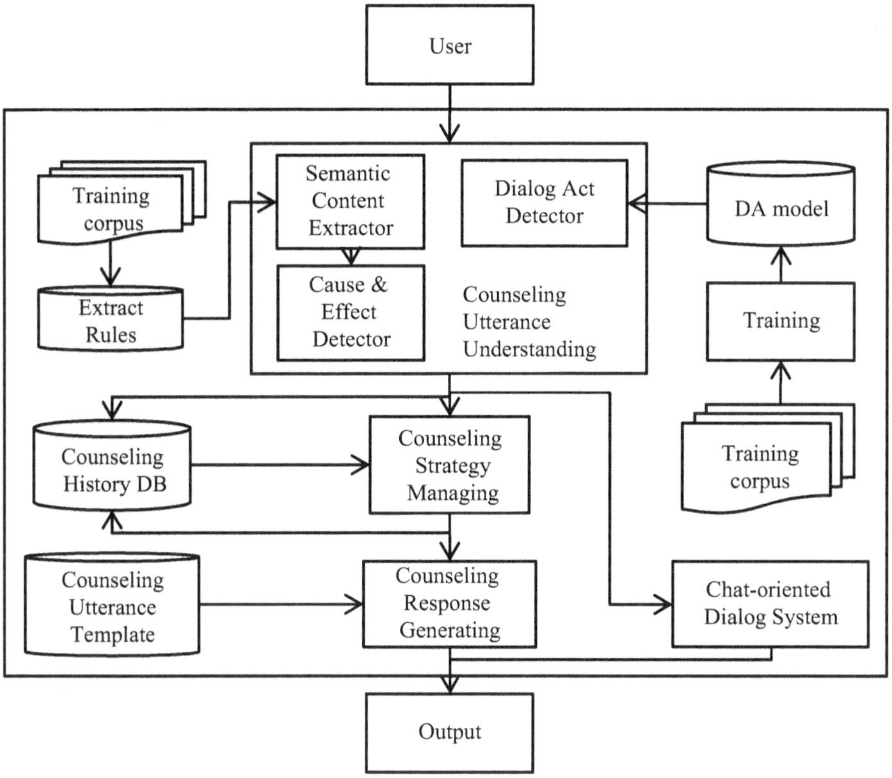

Fig. 6.1 System architecture

dialog system because we cannot generate a micro-counseling utterance. To extract semantic content, we use the dependency pattern matching method that is used in WOE^{parse} (Wu and Weld 2010). The dependency pattern is a partial dependency graph in which each node has a POS tag and each edge has a dependency label. Among those nodes, three nodes are marked as subject, predicate, and object. If a dependency pattern is found in the dependency graph of the user utterance, its corresponding subject, predicate, object phrases are extracted. We manually collected 360 dependency patterns from dependency graphs of the Movie-Dic corpus.

During a micro-counseling dialog, the system asks the user three types of questions: problem questions, reason questions, and emotion questions. Through the system questions, the system can detect a user utterance as the one that seeks counseling. For example, when a system asks the user about a problem, the user's answer is assumed to identify the problem.

Some user utterances can provide more than one type of counseling information. For example, "I feel sad because my dog died" includes two semantic contents: "I feel sad" is emotion information, and "my dog died" is problem information.

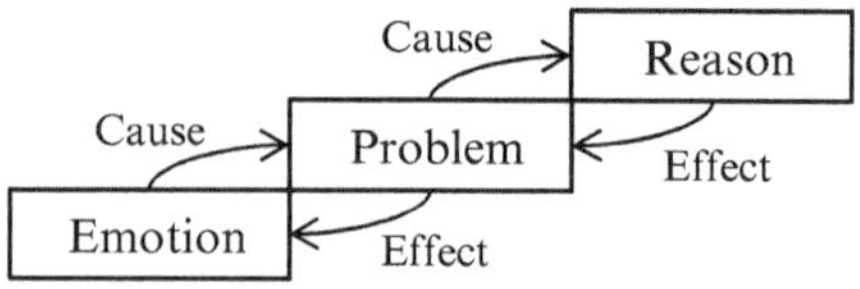

Fig. 6.2 Counseling information relationship

Table 6.3 Counseling strategy table

	Information in current user utterance			Information in dialog history		
Counseling strategy	Emotion	Problem	Reason	Emotion	Problem	Reason
Attending						
Paraphrasing	O					
		O				
			O			
Reflect feeling	X	O	O	O	X	X
	X	X	O	O	O	X
	O	X	X	X	O	O
		?			?	
Emotion question	X			X		
Problem question		X			X	
Reason question			X		O	X

'O': information should exist; 'X': information should not exist

To extract counseling information from this kind of double content utterance, we should consider the relationships between the types of counseling information (Fig. 6.2). We split the user utterance's semantic contents into cause and effect by comparing the location of semantic contents and classifying the conjunction. For example, in case of "I am sad because my dog died," "my dog died" causes "I am sad" because it is after the conjunction 'because.' We generated 14 rules to split semantic contents into cause and effect.

When the system asked a problem question, cause is assumed to be reason information and effect is assumed to be emotion information.

6.5.3 *Counseling Strategy Manager*

Our micro-counseling dialog system has four counseling strategies: attending, paraphrasing, reflect feeling, and question. We defined a counseling technique table that consists of strategies and the required conditions of each strategy (Table 6.3); the required conditions mean the existence of information in user's current utterance and dialog history. The system selects the best strategy based on counseling technique table.

Table 6.4 Counseling strategy example

Speaker	Utterance	Counseling strategy
System	Hello. How are you today?	Emotion question
User	I feel bad because I fought with my boyfriend.	–
System	You fought with your boyfriend. Why does it happen?	Paraphrase, reason question
User	He didn't remember my birthday.	–
System	I see. You feel bad because he didn't remember your birthday.	Paraphrase
User	That's right	–

- **Attending**: Attending utterances could follow after any kind of user utterances. So attending technique does not consider information extracted from the current or dialog history.
- **Paraphrasing**: Paraphrasing should follow a user utterance that includes at least one counseling information.
- **Reflecting**: Reflecting feelings should be used when information in current user utterance and information in dialog history includes whole counseling information.
- **Questioning**: Questioning techniques should be used to request information that has not been provided; emotion, problem, and reason. In case of reason questioning, problem information should exist in information in dialog history because reason should be asked after problem already known.

As an example of counseling dialog strategy (Table 6.4), the system asks an emotion question or a problem question at the beginning of dialog to induce the user speak. Questioning can be in a dialog turn with others.

6.5.4 Counseling Response Generation

Our system utterances are generated by using a counseling response template. We choose a system template by checking the counseling information extracted from the dialog and use extracted contents to fill slots in a counseling response template (Table 6.5). Each technique has its own templates, and each template has its own counseling information slots to fill.

Table 6.5 Counseling response template

System template	Counseling strategy
Oh I see.	Attending
You feel <eo>.	Paraphrasing
<es> <ep> <eo> because <ps> <pp> <po>.	Paraphrasing
You feel <eo> because <rs> <rp> <ro>.	Reflect feeling
Please tell me about your problem.	Problem question
How do you feel about <ps> did so?	Emotion question
Why did <ps> do so?	Reason question

Slots: <es> subject of emotion; <ep> predicate of emotion; <eo> object of emotion; <ps> subject of problem; <pp> predicate of problem; <po> object of problem; <rs> subject of reason; <rp> predicate of reason; <ro> object of reason

6.5.5 *Chat-Oriented Dialog System*

The chat-oriented dialog system can respond to any kind of user input sentence whether or not it is related to the counseling purpose. The system selects the most appropriate response from the chatting cues given the user input. This is based on the EBDM (Lee et al. 2009) framework; detailed description is beyond the scope of this paper. We only explain the example matching method. An example is a pair of a user-side sentence u and a system-side response s. We adopt a sentence similarity score with POS weights ($\mathrm{sim}_{\mathrm{POS}}$) to find the most appropriate responses as follows:

$$\mathrm{sim}_{\mathrm{POS}}(u, s) = \frac{2 \cdot |u \cap s|}{|u| + |s|}$$

The intersection is the set of words that occur in both sentences. When finding a matching word, coarse-grained POS tags and lemmatized words are used to ignore inflectional changes of the words. We also define POS weights and assign the word weight according to its POS. Finally, $|u|$, $|s|$, and $|u \cap s|$ are defined as the sum of all word weights in u, s, and $u \cap s$, respectively.

6.6 Experiment and Discussion

We first tested the performance of dialog act detection and semantic content extraction modules. Our fivefold cross-validation experiment test dataset includes a chatting corpus and a counseling corpus. The whole 11,840 utterances are labeled with dialog act and semantic contents that can generate a counseling response. Our experiment achieved >89 % statement dialog act detection performance and >95 % semantic content extraction performance as shown in Table 6.6.

Table 6.6 Dialog act and semantic content detection result

	Statement dialog-act detection	Semantic content extraction
Precision	88.9 %	97.4 %
Recall	89.6 %	92.7 %
F measure	89.3 %	95.0 %

Table 6.7 Experiment result ($p < 0.01$ for each question)

Question	Baseline	Proposed
System extracted appropriate information.	5.33	7.43
System understood my various utterances.	5.19	7.00
Information that system focused was appropriate.	5.90	7.19
System's dialog strategy was appropriate.	5.68	7.28
There was no interruption in my dialog.	6.43	9.19
I wanted to chat more with the system.	4.10	6.57

We recruited 16 volunteers to evaluate the effectiveness of the counseling information extraction method, the counseling strategy, and the chat back-off strategy. The baseline system for comparison is a previous counseling dialog system that uses 5W1H extraction. We gave 20 counseling situations to each user and asked them to talk to each system for a total of 30 min.

Each volunteer scored six evaluation questions on a scale of 1(low) to 10. To evaluate the CUU module based on semantic content extraction, the questions were asked users how much they were satisfied by the system's ability to understand their utterances. To assess the CSM module's counseling strategy, the questions were asked whether they were satisfied with its counseling strategy on the counseling information. To assess the back-off strategy we asked them to assess the relevance of its responses. Our system achieved a higher score overall than the baseline system (Table 6.7).

User satisfaction increased because the counseling information was extracted from various utterances. The redefined counseling information encouraged the user to interact intensively with the system. The chat-oriented back-off strategy increased overall satisfaction because it avoided interruption of dialogs.

6.7 Conclusion

We developed a counseling dialog system that extracts semantic counseling information, defines counseling information, and uses a chat-oriented dialog system as a back-off strategy. Because the counseling dialog system was developed for various user utterances, it can be used for other research in human–computer interaction such as development of health informatics and companions for seniors. Our future work is to improve our system to generate various system utterances that use additional micro-counseling techniques (Ivey et al. 2013).

Acknowledgments This work was partly supported by ICT R&D program of MSIP/IITP [10044508, Development of Non-Symbolic Approach-based Human-Like Self-Taught Learning Intelligence Technology] and National Research Foundation of Korean (NRF) [NRF-2014R1A2A1A01003041, Development of multi-party anticipatory knowledge-intensive natural language dialog system].

References

Banchs RE (2012) Movie-DiC: a movie dialogue corpus for research and development. In: Proceedings of the 50th annual meeting of the association for computational linguistics, Jeju, Republic of Korea, pp 203–207

Beger AL, Della Pietra SA, Della Pietra VJ (1996) A maximum entropy approach to natural language processing. Association for Computational Linguistics, pp 39–71

Evans DR, Hearn MT, Uhlemann MR, Ivey AE (2010) Essential interviewing, 8th edn. Cengage Learning

Han S, Lee K, Lee D, Lee GG (2013) Counseling dialog system with 5W1H extraction. In: Proceedings of the SIGDIAL2013 conference, Metz, France, pp 349–353

Ivey AE, Ivey MB, Zalaquett CP (2013) Intentional interviewing and counseling, 8th edn. Cengage Learning

Lee C, Jung S, Kim S, Lee GG (2009) Example-based dialog modeling for practical multi-domain dialog system. Speech Commun 51(5):466–484

Li Y, Bandar Z, McLean D, O'Shea J (2004) A method for measuring sentence similarity and its application to conversational agents. In: The 17th international FLAIRS conference, Florida, USA, pp 820–825

Meguro T, Minami Y, Higashinaka R, Dohsaka K (2013) Learning to control listening-oriented dialogue using partially observable Markov decision processes. ACM Trans Speech Lang Process 10(4), Article 15

Weizenbaum J (1966) ELIZA – a computer program for the study of natural language communication between man and machine. Commun Assoc Comput Mach 9:36–45

Wu F, Weld DS (2010) Open information extraction using Wikipedia. In: Proceedings of the 48th annual meeting of the association for computational linguistics, ACL'10, Morristown, NJ, USA, pp 118–127

Chapter 7
Users' Belief Awareness in Reinforcement Learning-Based Situated Human–Robot Dialogue Management

Emmanuel Ferreira, Grégoire Milliez, Fabrice Lefèvre, and Rachid Alami

Abstract Others can have a different perception of the world than ours. Understanding this divergence is an ability, known as perspective taking in developmental psychology, that humans exploit in daily social interactions. A recent trend in robotics aims at endowing robots with similar mental mechanisms. The goal then is to enable them to naturally and efficiently plan tasks and communicate about them. In this paper we address this challenge extending a state-of-the-art goal-oriented dialogue management framework, the Hidden Information State (HIS). The new version makes use of the robot's awareness of the users' belief in a reinforcement learning-based situated dialogue management optimisation procedure. Thus the proposed solution enables the system to cope not only with the communication ambiguities due to noisy channel but also with the possible misunderstandings due to some divergence among the beliefs of the robot and its interlocutor in a human–robot interaction (HRI) context. We show the relevance of the approach by comparing different handcrafted and learnt dialogue policies with and without divergent belief reasoning in an in-house pick–place–carry scenario by means of user trials in a simulated 3D environment.

Keywords Human-robot interaction • POMDP-based Dialogue Management • Reinforcement learning • Theory of mind

7.1 Introduction

When robots and humans share a common environment, previous works have shown how much enhancing the robot's perspective taking and intention detection abilities improves its understanding of the situation and leads to more appropriate and

---

E. Ferreira (✉) • F. Lefèvre
LIA/CERI, University of Avignon, Avignon, France
e-mail: emmanuel.ferreira@univ-avignon.fr; fabrice.lefevre@univ-avignon.fr

G. Milliez • R. Alami
CNRS LAAS - University of Toulouse, Toulouse, France
e-mail: gregoire.milliez@laas.fr; rachid.alami@laas.fr

G. Geunbae Lee et al. (eds.), *Natural Language Dialog Systems and Intelligent Assistants*, DOI 10.1007/978-3-319-19291-8_7

efficient task planning and interaction strategies (Breazeal et al. 2006, 2009; Milliez et al. 2014b). As part of the theory of mind, perspective taking is a widely studied ability in developmental literature. This broad term encompasses: (1) perceptual perspective taking, whereby human can understand that other people see the world differently, and (2) conceptual perspective taking, whereby humans can go further and attribute thoughts and feelings to other people (Baron-Cohen and Leslie 1985). Tversky et al. (1999) explain to what extent switching between perspectives rather than staying in an egocentric position can improve the overall dialogue efficiency in a situated context. Therefore, to make robots more socially competent, some research aims to endow robots with this ability. Among others, Breazeal et al. (2006) present a learning algorithm that takes into account information about a teacher's visual perspective in order to learn specific coloured buttons' activation/deactivation patterns, and Trafton et al. (2005) use both visual and spatial perspective taking to find out the referent indicated by a human partner. In the present study, we specifically focus on a false belief task as part of the conceptual perspective taking. Formulated in Wimmer and Perner (1983), this kind of task requires the ability to recognise that others can have beliefs about the world that differ from the observable reality. Breazeal et al. (2009) proposed one of the first human–robot implementations and proposed some more advanced goal recognition skills relying on this false belief detection. In Milliez et al. (2014b), a Spatial Reasoning and Knowledge component (SPARK) is presented to manage separate models for agent belief state and used to pass the Sally and Anne test (Baron-Cohen and Leslie 1985) on a robotic platform. This test is a standard instance of false belief task where an agent has to guess the belief state of another agent with a divergent belief mind state. The divergence in this case arises from modifications of the environment which one agent is unaware of and which are not directly observable, for instance displacement of objects hidden to this agent (behind another object for instance).

Considering this, to favour the human intention understanding and improve the overall dialogue strategy, we take benefit of the divergent belief management into the multimodal situated dialogue management problem. To do so, we rely on the Partially Observable Markov Decision Process (POMDP) framework. This latter is becoming a reference in the Spoken Dialogue System (SDS) field (Young et al. 2010; Thomson and Young 2010; Pinault and Lefèvre 2011) as well as in HRI context (Roy et al. 2000; Lucignano et al. 2013; Milliez et al. 2014a), due to its capacity to explicitly handle parts of the inherent uncertainty of the information which the system (the robot) has to deal with (erroneous speech recogniser, falsely recognised gestures, etc.). In the POMDP setup, the agent maintains a distribution over possible dialogue states, the belief state, all along the dialogue course and interacts with its perceived environment using a reinforcement learning (RL) algorithm so as to maximise some expected cumulative discounted reward (Sutton and Barto 1998). So our goal here is to introduce the divergence notion into the belief state tracking and add some means to deal with it in the control part.

The remainder of the paper is organised as follows. Section 7.2 gives some details about how an agent knowledge model can be maintained in a robotic system; in Sect. 7.3 our extension of a state-of-art goal-oriented POMDP dialogue management framework, the Hidden Information State (HIS), is presented to take into account users' beliefs state; in Sect. 7.4 the proposed pick–place–carry false belief scenario used to exemplify the benefit of both taking account of the perspective taking ability and its integration in a machine learning scheme is introduced. In the same section, the current system architecture and the experimental setup employed are given. The user trial results obtained with a learnt and a handcrafted belief-aware system are compared in Sect. 7.5 with systems lacking perspective taking ability. Finally, in Sect. 7.6 we discuss some conclusions and give some perspectives.

7.2 Agent Knowledge Management

As mentioned in the introduction, the spatial reasoning framework SPARK is used for situation assessment and spatial reasoning. We will briefly recap here how it works, for further details please refer to Milliez et al. (2014b). In our system, the robot collects data about three different entities to virtually model its environment: objects, humans and *proprioceptions* (its own position, posture, etc.). Concerning objects, a model of the environment is loaded at startup to obtain the positions of static objects (e.g. walls, furnitures, etc.). Other objects (e.g. mug, tape, etc.) are considered as movable. Their positions are gathered using the robot's stereo vision. Posture sensors, such as Kinect, are used to obtain the position of humans. These perception data allow the system to use the generated virtual model for further spatial-temporal reasoning. As an example, the system can reason on why an object is not perceived any more by a participant and decide to keep its last known position if it recognizes a situation of occlusion, or remove the object from its model if there is none.

Figure 7.1a shows a field experiment with the virtual environment built by the system from the perception data collected and enriched by the spatial reasoner. The latter component is also used to generate facts about the objects relative position and agents' affordances. The relative positions such as *isIn*, *isNextTo*, *isOn* are used not only for multimodal dialogue management as a way to solve referents in users' utterances, but also for a more natural dialogue description of the objects position in the robot's responses. Agents' affordances come from their ability to perceive and

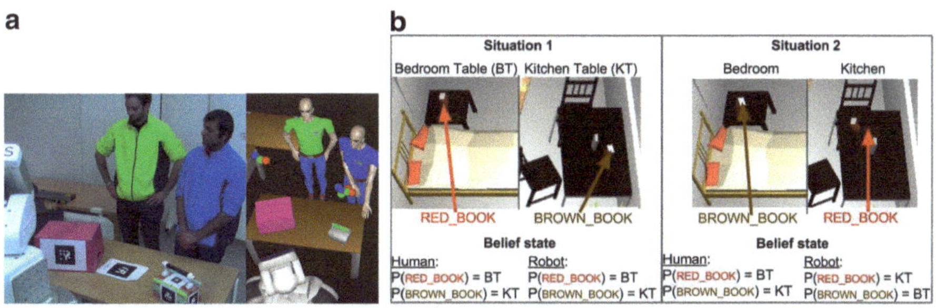

Fig. 7.1 (**a**) Real users in front of the robot (*left*) and the virtual representation built by the system (*right*). (**b**) Divergent belief example with belief state

reach objects. The robot is calculating its own capability of perception according to the actual data it gets from the object position and recognition modules. For reachability, the robot computes if it is able to reach the object with its grasping joints. To compute the human's affordances the robot applies its perspective taking ability. In other words, the robot has to estimate what is visible and reachable for the human according to her current position. For visibility, it computes which objects are present in a cone, emerging from human's head. If the object can be directly linked to the human's head with no obstacle and if it is in the field of the view cone, then it is assumed that the human sees the object and hence has knowledge of its true position. If an obstacle is occluding the object, then it won't be visible for the human. Concerning the reachability, a threshold of one meter is used to determine if the human can reach an object or not.

The facts generation feature allows the robot to get the information about the environment, its own affordances and the human's affordances. In daily life, humans get the information about the environment through perception and dialogue. Using the perspective taking abilities of our robot, we can compute a model of each human's belief state according to what she perceived or what the robot has told her about the environment. Then two different models of the world are considered: one for the world state from the robot perception and reasoning and one for each human's belief state (computed by the robot according to what the human perceived). Each of these models is independent and logically consistent. In some cases, the robot and the human models of the environment can diverge. As an example, if an object *O* has a property *P* with a value *A*, if *P*'s value changed to *B* and the human had no way to perceive it when it occurred, the robot will have the value *B* in its model ($P(O) = B$) while the human will still have the value *A* for the property *P* ($P(O) = A$). This value shouldn't be updated in the human model until the human is actually able to perceive this change or until the robot informs him. In our scenario, this reasoning is applied to the position property.

We introduce here an example of false belief situation (Fig. 7.1b). A human sees a red book (RED_BOOK) on the bedside table *BT*. She will then have this property in his belief state: P(RED_BOOK) = *BT*. Now, while this human is away (has no perception of *BT*), the book is swapped with another brown one (BROWN_BOOK) from the kitchen table *KT*. In this example, the robot explores the environment and is aware of the new position values. The human will keep this belief until she gets a new information on the current position of RED_BOOK. This could come from actually seeing RED_BOOK on the position *KT* or seeing that RED_BOOK is not any more in *BT* (in which case the position property value will be updated to an *unknown* value). Another way to update this value is for the robot to explicitly inform the user of the new position.

In our system we mainly focused on position properties but this reasoning could be straightforwardly extended to other properties such as who manipulated an object, its content, temperature, etc. Obviously if this setup generalises quite easily to false beliefs about individual properties of elements of the world, more complex divergence configurations that might arise in daily interactions, for instance due to prior individual knowledge, still remain out of range and should be addressed by future complementary works.

7.3 Belief Aware Multimodal Dialogue Management

As mentioned earlier, an important aspect of the approach is to base our user belief state management on the POMDP framework (Kaelbling et al. 1998). It is a generalisation of the fully observable Markov Decision Process (MDP), which was first employed to determine an optimal mapping between situations (dialogue states) and actions for the dialogue management problem in Levin et al. (1997). We try hereafter to recall some of the principles of this approach pertaining to the modifications that will be introduced. More comprehensive descriptions should be sought in the cited papers. This framework maintains a probability distribution over dialogue states, called belief states, assuming the true one is unobservable. By doing so, it explicitly handles parts of the inherent uncertainty on the information conveyed inside the Dialogue Manager (DM) (e.g. error prone speech recognition and understanding processes). Thus, POMDP can be cast as a continuous space MDP. The latter is a tuple $< B, A, T, R, \gamma >$, where B is the belief state space (continuous), A is the discrete action space, T is a set of Markovian transition probabilities, R is the immediate reward function, $R : B \times A \times B \rightarrow \Re$ and $\gamma \in [0, 1]$ the discount factor (discounting long-term rewards). The environment evolves at each time step t to a belief state b_t and the agent picks an action a_t according to policy mapping belief states to actions, $\pi : B \rightarrow A$. Then the belief state changes to b_{t+1} according to the Markovian transition probability $b_{t+1} \sim T(.|b_t, a_t)$ and, following this, the agent received a reward $r_t = R(b_t, a_t, b_{t+1})$ from the environment. The overall problem of this continuous MDP is to derive an optimal policy maximising the reward expectation. Typically the averaged discounted sum over a potentially infinite horizon is used, $\sum_{t=0}^{\infty} \gamma^t r_t$. Thus, for a given policy and start belief state b, this quantity is called the value function: $V^{\pi}(b) = E[\sum_{t \geq 0} \gamma^t r_t | b_0 = b, \pi] \in \Re^B$. V^* corresponds to the value function of any optimal policy π^*. The Q-function may be defined as an alternative to the value function. It adds a degree of freedom on the first selected action, $Q^{\pi}(b, a) = E[\sum_{t \geq 0} \gamma^t r_t | b_0 = b, a_0 = a, \pi] \in \Re^{B \times A}$, Q^* corresponds to the action-value function of any optimal policy π^*. If it is known, an optimal policy can be directly computed by being greedy according to Q^* , $\pi^*(b) = \arg\max_a Q^*(b, a) \forall b \in B$.

However, real-world POMDP problems are often intractable due to their dimensionality (large belief state and action spaces). Among other techniques, the HIS model (Young et al. 2010) circumvents this scaling problem for dialogue management by the use of two main principles. First, it factors the dialogue state into three components: the user goal, the dialogue history and the last user act (see Fig. 7.2). The possible user goals are then grouped together into *partitions* on the assumption that all goals from the same partition are equally probable. These partitions are built using the dependencies defined in a domain-specific ontology and the information extracted all along the dialogue from both the user and the system communicative acts. In the standard HIS model, each partition is linked to matching database entities based on its static and dynamic properties that correspond to the current state of the world (e.g. colour of an object vs spatial relations like *isOn*).

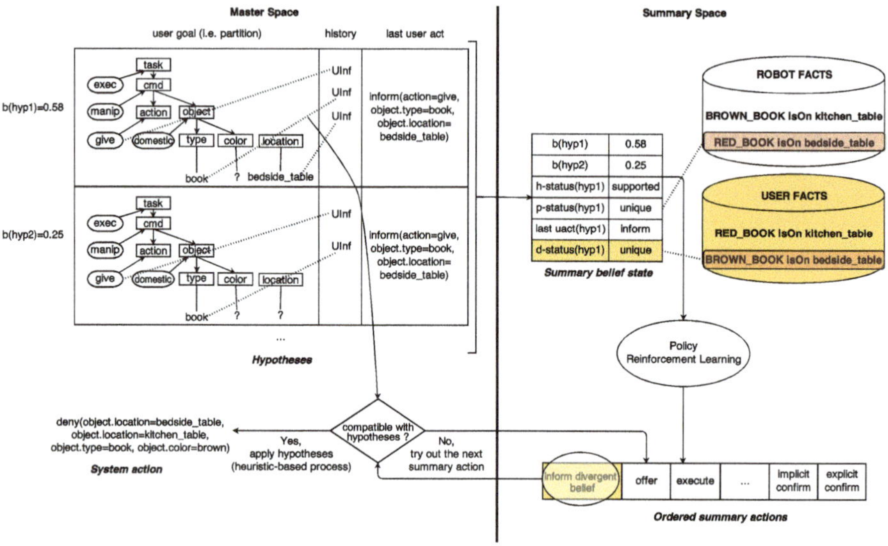

Fig. 7.2 Overview of the HIS extension to take into account divergent belief

The combination of a partition, the associated dialogue history, which corresponds here to a finite state machine that keeps track of the grounding status for each convoyed piece of information (e.g. informed or grounded by the user), and a possible last user action forms a dialogue state hypothesis. A probability distribution $b(\text{hyp})$ over the most likely hypotheses is maintained during the dialogue and this distribution constitutes the POMDP's belief state. Second, HIS maps both the belief space (hypotheses) and the action space into a much reduced summary space where RL algorithms are tractable. The summary state space is the compound of two continuous and three discrete values. Continuous values are the probabilities of the two-first hypotheses $b(\text{hyp1})$ and $b(\text{hyp2})$ while the discrete ones, extracted from the top hypothesis, are the type of the last user act (noted *last uact*), a partition status (noted *p-status*) database matching status related to the corresponding goal and a history status (noted *h-status*). Likewise system dialogue acts are simplified in a dozen of summary actions like *offer*, *execute*, *explicit-confirm* and *request*. Once the summary actions are ordered by their $Q(b, a)$ scores in descending order by the policy, a handcrafted process checks if the best scored action is compatible with the current set of hypotheses (e.g. for the *confirm* summary act this compatibility test consists in checking if there is something to confirm in the top hypothesis). If they are compatible, a heuristic-based method maps this action back to the master space as the next system response. If not, the process is pursued using the next best scored summary action until a possible action is found.

The standard HIS framework can properly handle misunderstandings due to noise in the communicative channel. However, misunderstandings can also be introduced in cases where the user has false beliefs, impacting negatively her communicative acts. HIS has no dedicated mechanism to deal with such a situation and so it should react as in front of a classical uncertainty by asking the user to confirm hypotheses until the request can match the reality, although it could have been resolved since the first turn. Therefore having an appropriate mechanism should improve the quality and efficiency of the dialogue, preventing user to pursue her goal with an erroneous statement.

So, as illustrated in Fig. 7.2 and highlighted with the orange items, we propose to extend the summary belief state with an additional status, the *divergent belief* status (noted *d-status*), and an additional summary action, *inform divergent belief*. The *d-status* is employed to trigger the presence of false belief situations by matching the top partition with user facts compiled by the system (see Sect. 7.2) and as such trying to highlight some divergences between the user and the robot points of view. Both the user and the robot facts (from the belief models, not to be mistaken with the belief state related to the dialogue representation) are considered as part of the dynamic knowledge resource and are maintained independently of the internal state of the system with the techniques described in Sect. 7.2. Here we can observe in Fig. 7.2 that the top partition is about a book located on the bedside table. In the robot model of the world (i.e. robot facts) this book is identified as a unique entity, RED_BOOK, and *p-status* is set to *unique* accordingly. However, in the user model it is identified as BROWN_BOOK. This situation can be considered as divergent and *p-status* is set to *unique* too because there is one possible object that corresponds to that description in the user model. In this preliminary study *d-status* can only be *unique* or *non-unique*. Further studies may consider more complex cases. The new summary action is employed for appropriate resolution and removal of the divergence. The (real) communicative acts associated to this (generic) action rely on expert design. In this first version, if this action is compatible with the current hypotheses and thus picked up by the system, it explicitly informs the user of the presence and the nature of the divergence. To do so, the system uses a *deny* dialogue act to inform the user about the existence of a divergent point of view and let the user agree on the updated information. Consequently, the user may pursue its original goal with the correct property instead of the obsolete one. This process is also illustrated in Fig. 7.2 when the *inform divergent belief* action is mapped back to the master space.

7.4 Scenario and Experimental Setup

In order to illustrate the robot's ability to deal with user's perspective, an adapted pick–place–carry scenario is used as test-bed. The robot and the user are in a virtual flat with three rooms, in which there are different kinds of objects varying in terms of colour, type and position (e.g. blue mug on the kitchen table, red book on the living room table, etc.). The user interacts with the robot using unconstrained speech (Large Vocabulary Speech Recognition) and pointing gestures to ask the robot to perform some specific object manipulation tasks (e.g. move the blue mug from

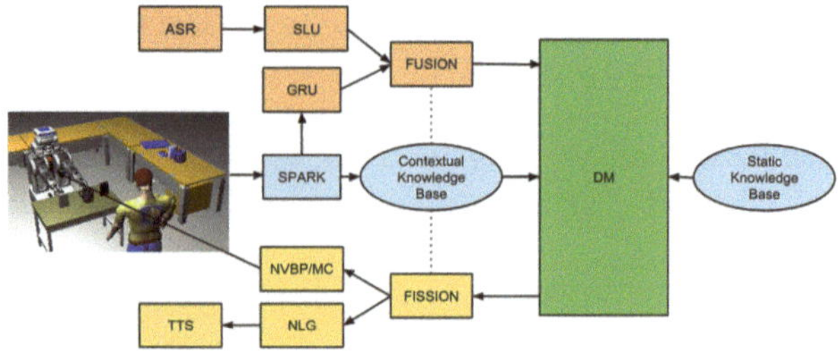

Fig. 7.3 Architecture of the multimodal and situated dialogue system

the living room table to the kitchen table). The multimodal dialogue is used to solve ambiguities and to request missing information until task completion (i.e. full command execution) or failure (i.e. explicit user disengagement or wrong command execution). In this study, we specifically focus on tasks where divergent beliefs are prone to be generated as in the Sally and Anne test: a previous interaction has led the user to think that a specific object O is located at A which is out of her view, and an event has changed the object position from A to B without user's awareness. For example, a change performed by another user (or by the robot) without the presence of the first one. Thereby, if the user currently wants to perform a manipulation involving O she may do so using her own believed value (A) of the position property in her communicative act.

Concerning the simulation, the setup of Milliez et al. (2014a) is applied to enable a rich multimodal HRI. Thus, the open-source robotics simulator MORSE (Echeverria et al. 2011) is used which provides a realistic rendering through the Blender Game Engine, a wide range support of middleware (e.g. ROS, YARP), and proposes reliable implementations of realistic sensors and actuators which ease the integration on real robotic platforms. It also provides the operator with an immersive control of a virtual human avatar in terms of displacement, gaze and interactions on the environment, such as object manipulation (e.g. grasp/release an object). This simulator is tightly coupled with the multimodal dialogue system, with the overall architecture given in Fig. 7.3.

In the chosen architecture, the Google Web Speech API[1] for Automatic Speech Recognition (ASR) is combined with a custom-defined grammar parser for Spoken Language Understanding (SLU). The spatial reasoning module, SPARK, is responsible for both detecting the user gestures and generating the per-agent spatial facts (see Sect. 7.2) used to dynamically feed the contextual knowledge base and allowing the robot to reason over different perspectives of the world. Furthermore, we also make use of a static knowledge base containing the list of all available objects (even those not perceived) and their related static properties (e.g. colour). The Gesture Recognition and Understanding (GRU) module catches the gesture-events generated by SPARK during the course of the interaction. Then, a rule-based fusion engine, close to the one presented in Holzapfel et al. (2004), temporally aligns

[1]https://www.google.com/intl/en/chrome/demos/speech.html.

the monomodal inputs (speech and gesture) and merges them to convey the list of possible fused inputs to the POMDP-based DM, with speech considered as the primary modality.

The DM implements the extended HIS framework described in Sect. 7.3. For the reinforcement learning setup, the sample-efficient KTD-SARSA RL algorithm (Daubigney et al. 2012) in combination with the Bonus Greedy exploration scheme enables online learning of dialogue strategy from scratch, as in Ferreira and Lefevre (2013a). A reward function is defined to penalise the DM by -1 for each dialogue turn and give it a $+20$ if the right command is performed at the end of the interaction, 0 otherwise. To convey the DM action back to the user, a rule-based fission module is employed that splits the high-level DM decision into verbal and non-verbal actions. The robot speech outputs are generated by chaining a template-based Natural Language Generation (NLG) module, which converts the sequence of concepts into text, to a Text-To-Speech (TTS) component based on the commercial Acapela TTS system.[2] A Non-verbal Behaviour Planning and Motor Control (NVBP/MC) module produces robot postures and gestures by translating the non-verbal actions into a sequence of abstract actions such as *grasp*, *moveTo*, *release* which are then executed in the simulated environment.

In this study we intend to assess the benefit of introducing the divergent belief management into the multimodal situated dialogue management problem. Thereby, the scenarios of interest require some situations of divergent beliefs between the user and the robot. In real setup those scenarios often need a long-term interaction context tracking. To bypass this time-consuming process in our evaluation setup, we directly propose a corrupted goal to the user at the beginning of her interaction. So, a false belief about the location value was automatically added concerning an object not visible from the human point of view. Although the situation is artificially generated, the same behaviour can be obtained with the spatial reasoner if the robot performs an action in self-decision mode or if another human corrupts the scene. Thereby, this setup was used to evaluate the robot's ability to deal with both classical (CLASSIC) and false belief (FB) object manipulation tasks. To do so, we compare the belief-aware learnt system performance (noted BA-LEARNT hereafter) to a handcrafted one (noted BA-HDC), and with two other similar systems with no perspective taking ability (noted LEARNT and HDC, respectively). The handcrafted policies make use of expert rules based on the information provided by the summary state to pick the next action to perform (deterministic). They are not considered as the best possible handcrafted policies but as robust enough to manage correctly an interaction with real users. The learnt policies were trained in an online learning setting using a small set of 2 expert users which first performed 40 dialogues without FB tasks and 20 more as a method-specific adaptation (LEARNT with CLASSIC tasks vs BA-LEARNT with FB tasks). In former works we have shown the possibility to learn efficient policies with few tens of dialogue samples, due to expert users' better tolerance to poor initial performance combined with more consistent behaviours during interactions (Ferreira and Lefèvre 2013b).

[2]http://www.acapela-group.com/index.html.

In the evaluation setup, ten dialogues for the four proposed system configurations (the learnt policies were configured to act greedily according to the value function) were recorded from six distinct subjects (two females and four males, around 25 years old on average) who interacted with all configurations (within-subjects study), so 240 dialogues in total. Thirty percent of the performed dialogues involve FB tasks. No user had knowledge of the current system configurations and they were proposed in random order to avoid any prior effect. At the end of each interaction, users evaluated the system in terms of task completion with an online questionnaire.

7.5 Results

Table 7.1 is populated with the performance obtained by the four system configurations discussed above considering CLASSIC and FB tasks. These results are first given in terms of mean discounted cumulative rewards (Avg.R). According to the reward function definition, this metric expresses in a single real value the two variables of improvement, namely the success rate (accuracy) and the number of turns until dialogue end (time efficiency). However, both metrics are also presented for convenience. The results in Table 7.1 were gathered in test condition where no exploration of the RL method is allowed. Thus, they basically consist of a mere average over the 60 performed dialogues for each method and metric.

The differences observed between the LEARNT/BA-LEARNT and the HDC/BA-HDC on the overall performance (row ALL) show the interest of considering RL methods rather than handcrafted policies. Indeed, only 60 training dialogues are enough to outperform both handcrafted solutions. On CLASSIC tasks the performance between LEARNT and BA-LEARNT as well as between HDC and BA-HDC must be considered similar. Thus, the divergent belief resolution mechanism doesn't seem to impact the dialogue management when divergent belief situations do not appear. For BA-HDC this statement could be expected (in lack of false belief, the rules are the same as HDC). However for BA-LEARNT the tested policy is learnt and the action assignment process is optimised with an additional degree of complexity (larger state/action space than in LEARNT), so a loss could have been observed. The performances between LEARNT and BA-LEARNT and respectively between HDC and BA-HDC on FB tasks appear in favour of the BA-systems (both show a higher success rate and a slightly more time efficient dialogue management process—average gain of 1 turn). However the quantitative comparison between the system configurations is not ensured to be relevant due to the relatively high confidence interval on considered metrics (e.g. success rate confidence interval for row FB is around 0.2 for all system configurations). Two main reasons account for this status quo. First, a limited amount of observations involving the different system configurations (due to experimental cost). Second, the expected marginal gain in terms of the considered metrics. Indeed, the current system is learnt on some overall task completion and efficiency criterion. However solving divergent belief situations in a pick and place scenario cannot be considered

Table 7.1 System performance on classic (CLASSIC), false belief (FB) and all (ALL) tasks in terms of average cumulative discounted reward (Avg.R), average dialogue length in terms of system turns (Length) and average success rate (SuccR)

	HDC			BA-HDC			LEARNT			BA-LEARNT		
TASK	Avg.R	Length	SuccR	Avg.R	Length	SuccR	Avg.R	Length	SuccR	Avg.R	Length	SuccR
CLASSIC	14.33	4.81	0.85	14.28	4.86	0.86	17.62	2.95	0.93	17.69	2.88	0.93
FB	9.78	6.67	0.72	13.05	5.61	0.83	12.72	5.94	0.83	13.89	4.78	0.83
ALL	12.97	5.36	0.82	13.92	5.08	0.85	16.15	3.85	0.9	16.55	3.45	0.9

Table 7.2 Dialogue examples with (a) and without (b) divergent belief reasoning in the case of an unknown (from the user's point of view) interchange between a red and a brown book

R_1:	Can I help you?	U_1:	Bring me the book on my bedside table
		R_2:	Do you want the red one?
R_2:	**The brown book is not on the bedside table, it has been moved to the kitchen**	U_2:	No, the brown book
		R_3:	There is no brown book in your bedroom but there is one in the kitchen
U_2:	Ok, bring it to me	U_3:	Are you sure? Well, bring me that one
		R_4:	I will give you the brown book that is on the kitchen table
R_3:	I will give you the brown book that is on the kitchen table		
(a)		(b)	

a critical factor influencing this criterion greatly but just a way to cope with an additional (not dominant) degree of uncertainty and to improve user experience and naturalness of the interaction with the embodied agent.

To have better insights on what the main differences between the four dialogue strategies are we also performed a qualitative study. In this study we precisely identify the behavioural differences due to introducing an FB handling mechanism in a learning setup. Overall, it is observed that confirmation acts (e.g. confirm, offer) are more accurate and less frequent for the two learnt methods. For instance, when the learnt systems are confident on the top object manipulation hypothesis they predominantly performed the command directly rather than trying to check its validity further as in the handcrafted versions. In Table 7.2 two dialogue samples extracted from the evaluation dataset illustrate the differences between non-BA and BA dialogue management on the same FB task (here a red book was interchanged with a brown one). If the belief divergence problem is not explicitly taken into account (as in (a)) the DM can be constrained to deal with an additional level of misunderstanding (see (b) from R_2 to U_3). We can also see in (b) that the non-BA system was able to succeed FB tasks (explaining the relatively high LEARNT performance on FB tasks). Indeed, if the object is clearly identified by the user (e.g. colour and type) the system can release the constraint of the false position and thus is able to make an offer on (execute) the "corrected" form of the command involving the true object position. Concerning the main differences between BA-LEARNT and BA-HDC, we observed a less systematic usage of the *inform divergent belief* act in the learnt case. BA-LEARNT first tries to reach a high confidence on the true presence of the object involved in the belief divergence in the user goal. Furthermore, BA-LEARNT, like LEARNT, has learnt alternative mechanisms to fulfil FB tasks such as direct execution of the user command (which also avoids misunderstanding) when the convoyed piece of information seems to be sufficient to identify the object.

7.6 Conclusion

In this paper, we described how a user belief real-time tracking framework can be used along with a multimodal POMDP-based dialogue management. The evaluation of the proposed method with real users confirms that this additional information helps to achieve more efficient and natural task planning (and does not harm handling of normal situations). Our next step will be to integrate the multimodal dialogue system on the robot and carry out evaluations in real setting to uphold our claims in a fully realistic configuration.

Acknowledgements This work has been partly supported by the French National Research Agency (ANR) under project reference ANR-12-CORD-0021 MaRDi.

References

Baron-Cohen S, Leslie AM, Frith U (1985) Does the autistic child have a 'theory of mind'? Cognition 21(1):37–46

Breazeal C, Berlin M, Brooks A, Gray J, Thomaz A (2006) Using perspective taking to learn from ambiguous demonstrations. In: Robotics and autonomous systems

Breazeal C, Gray J, Berlin M (2009) An embodied cognition approach to mindreading skills for socially intelligent robots. Int J Robot Res 28:656

Daubigney L, Geist M, Chandramohan S, Pietquin O (2012) A comprehensive reinforcement learning framework for dialogue management optimization. J Sel Top Sign Process 6(8):891–902

Echeverria G, Lassabe N, Degroote A, Lemaignan S (2011) Modular open robots simulation engine: Morse. In: ICRA

Ferreira E, Lefevre F (2013a) Expert-based reward shaping and exploration scheme for boosting policy learning of dialogue management. In: ASRU

Ferreira E, Lefèvre F (2013b) Social signal and user adaptation in reinforcement learning-based dialogue management. In: Proceedings of the 2nd MLIS workshop. ACM, New York, pp 61–69

Holzapfel H, Nickel K, Stiefelhagen R (2004) Implementation and evaluation of a constraint-based multimodal fusion system for speech and 3d pointing gestures. In: ICMI

Kaelbling L, Littman M, Cassandra A (1998) Planning and acting in partially observable stochastic domains. Artif Intell J 101(1–2):99–134

Levin E, Pieraccini R, Eckert W (1997) Learning dialogue strategies within the Markov decision process framework. In: ASRU

Lucignano L, Cutugno F, Rossi S, Finzi A (2013) A dialogue system for multimodal human-robot interaction. In: ICMI

Milliez G, Ferreira E, Fiore M, Alami R, Lefèvre F (2014a) Simulating human robot interaction for dialogue learning. In: SIMPAR, pp 62–73

Milliez G, Warnier M, Clodic A, Alami R (2014b) A framework for endowing interactive robot with reasoning capabilities about perspective-taking and belief management. In: ISRHIC

Pinault F, Lefèvre F (2011) Unsupervised clustering of probability distributions of semantic graphs for pomdp based spoken dialogue systems with summary space. In: KRPDS

Roy N, Pineau J, Thrun S (2000) Spoken dialogue management using probabilistic reasoning. In: ACL

Sutton R, Barto A (1998) Reinforcement learning: an introduction. IEEE Trans Neural Netw 9(5):1054–1054

Thomson B, Young S (2010) Bayesian update of dialogue state: a pomdp framework for spoken dialogue systems. Comput Speech Lang 24(4):562–588

Trafton J, Cassimatis N, Bugajska M, Brock D, Mintz F, Schultz A (2005) Enabling effective human-robot interaction using perspective-taking in robots. IEEE Trans Syst Man Cybern 35(4):460–470

Tversky B, Lee P, Mainwaring S (1999) Why do speakers mix perspectives? Spat Cogn Comput 1(4):399–412

Wimmer H, Perner J (1983) Beliefs about beliefs: representation and constraining function of wrong beliefs in young children's understanding of deception. Cognition 13(1):103–128

Young S, Gašić M, Keizer S, Mairesse F, Schatzmann J, Thomson B, Yu K (2010) The hidden information state model: a practical framework for pomdp-based spoken dialogue management. Comput Speech Lang 24(2):150–174

Chapter 8
Scalable Summary-State POMDP Hybrid Dialog System for Multiple Goal Drifting Requests and Massive Slot Entity Instances

Sangjun Koo, Seonghan Ryu, Kyusong Lee, and Gary Geunbae Lee

Abstract One of the main problems with partially observable Markov decision process (POMDP) in development of spoken dialog system (SDS) is lack of scalability. In development of an SDS with electronic program guide (EPG) domain, we devised a POMDP approach which is operated with summary spaces to respond accurately to multiple drifting goals and massive numbers of slot entities. The main point of the proposed approach is to introduce a hybrid architecture that is implemented by a meta-action selector and a service provider. A trained POMDP policy was used to select meta-actions. The selected meta-actions were transformed to the system action in the service provider, which is implemented with the given system action model. By using this architecture, various system actions could be elicited with reduced complexity in the dialog process. We trained the system with the specified simulator and observed its behavior with learning curves in the Korean EPG domain. The convergence of learning curve implies the feasibility of our approach in commercial EPG domain SDS.

Keywords Partially observable Markov decision process • Meta-action selector • Service provider

8.1 Introduction

A main challenge in development of spoken dialog systems (SDSs) is to ensure a certain level of accuracy of system responses to ambiguous user input. The input can be affected by several factors including classification errors committed by automatic speech recognition (ASR)/spoken language understanding (SLU).

S. Koo (✉) • S. Ryu • K. Lee • G. Geunbae Lee
Pohang University of Science and Technology, San 31, Pohang, Republic of Korea
e-mail: giantpanda@postech.ac.kr; ryush@postech.ac.kr; kyusonglee@postech.ac.kr; gblee@postech.ac.kr

G. Geunbae Lee et al. (eds.), *Natural Language Dialog Systems and Intelligent Assistants*, DOI 10.1007/978-3-319-19291-8_8

Construction of an SDS that can respond to ambiguous input can be modeled as a decision-making problem under uncertainty. One of the most widely used decision-making techniques is the partially observable Markov decision process (POMDP) framework which is derived from the Markov decision process (MDP) (Bellman 1957). The POMDP framework is suitable for designing SDSs (Bui et al. 2006; Gasic et al. 2013; Roy et al. 2000; Young et al. 2010; Zhang et al. 2001) because it allows SDS designers to elicit mathematical models of user behavior and uses automated stochastic process to train SDSs (Bui et al. 2009).

However, several problems obstruct the use of the POMDP framework in SDSs. One of the major problems is lack of scalability: the number of POMDP states increases exponentially as the number of named entities increases; this trend results in complexity that is too high for a real-time SDS (Bui et al. 2009). Two major trends of research have been conducted to solve this problem (Young et al. 2010): (1) factoring slot information into substates to reduce the number of POMDP states (Young et al. 2007) and (2) constructing partitions of given dialog states to prune out irrelevant states (Williams and Young 2007; Young 2006; Young et al. 2010).

Both approaches are based on the hypothesis that given initial goals of users do not easily change, and that if the goals change, their alterations can be described using simple rules. This assumption is valid in domains such as the tourist information guide (TIG), in which the intention of a user can be represented as a single specific goal. However, when we designed an SDS for the electronic program guide (EPG) domain, we observed that the hypothesis is not effective as they are in the TIG domain for three reasons: (1) Multiple drifting goals usually appear in a single dialog session, since most EPG dialogs consist of a series of independent requests and users do not have a specific goal in their mind. (2) The number of slot entity instances in the EPG domain is much larger than in the TIG domain. Entity slots in EPG typically cover open-category values including titles of TV programs and movies. (3) The dialog length is shorter in the EPG domain than in the TIG domain: most of the user requests are processed in a single request session.

In development of the SDS for EPG domain, we designed an architecture that is specified to operate with multiple drifting goals. The main idea of the proposed SDS is to separate overall dialog management (DM) into two components: meta-action selector and service provider. The meta-action selector was implemented with the POMDP framework to select meta-actions from given belief states; the service provider was implemented as a rule-based DM to select system actions for the user. The partition between two components reduces the number of dialog states, thereby reducing the complexity of the dialog process.

8.2 Overall Architecture

The main architecture (Fig. 8.1) consists of three subcomponents: tracker, meta-action selector, and service provider.

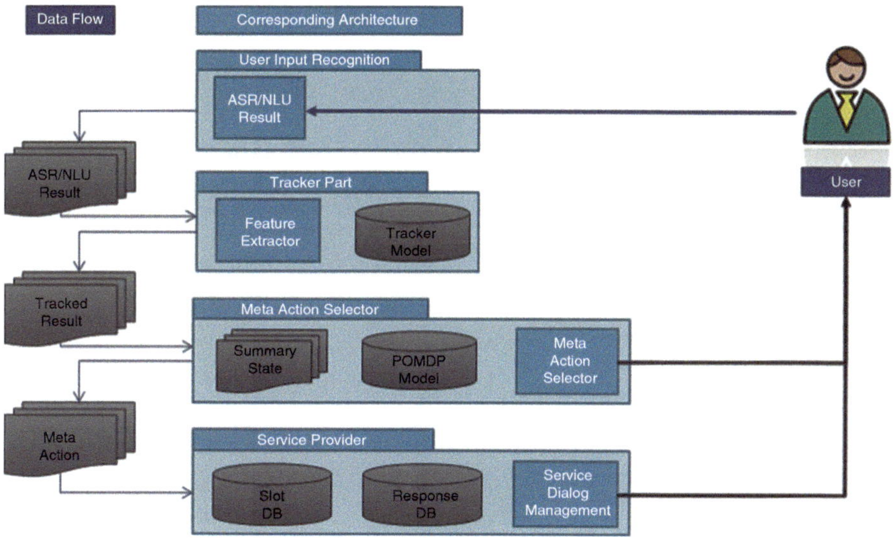

Fig. 8.1 Architecture of proposed SDS for EPG domain

Recognized user input from ASR/SLU is interpreted as a probability distribution of user intention slots. The dialog system considers this probability distribution to be an observation. *Tracker* tracks belief states of each entity slot individually and summarizes them into summary states. Summary states are compressed forms of original dialog states and are introduced to guide selection of POMDP actions. *Meta-action selector* uses a trained POMDP policy model to select meta-actions with tracked summary states. These meta-actions include "submit to service provider," "confirm the value of slot x," "request that the user provide the value of slot x." If the "submit" action is selected by the meta-action selector, *service provider* elicits the corresponding system action to provide the appropriate service to the user.

8.3 Tracking of Summary States

Let $o \in O$ be an observation of user input, let $b \in B$ be the internal state of given dialog system, and let $a \in A$ be system output. The process of tracking summary states can be divided into two subtasks: (1) Updating internal belief state with the given previous state, the user input and system output. This subtask can be represented as a conditional probability form $P(b'|b, a, o)$. (2) Constructing summary states from original internal belief states. This subtask can be represented as a mapping function $\phi(b') : B \to \bar{B}$.

8.3.1 Update of Goal States

A single tracker cannot easily track overall belief states. For example, 10 slots with 100 slot values each may yield as a total of 100^{10} states, which are unlikely to be tracked within reasonable time constraints. However, tracking each slot independently (Thomson and Young 2010) is a reasonable strategy. Without loss of generality, suppose a given system includes slots x and y. Then the belief probability $b : b(s)$ can be represented as

$$b(s) = b(s_x)b(s_y). \tag{8.1}$$

By indexing each entity slot $i = 1, \ldots, N$, marginal slot states can be stated as $b_i : b(s_i)$. Belief states for user intention should also be tracked. We introduced the concept of slot *UI* with belief states b_{UI} to represent the value of the user intention slot. The joint belief probability $b(s_{\mathrm{UI}}, s_i, s_j)$ can be represented as

$$b(s_{\mathrm{UI}}, s_i, s_j) = b(s_{\mathrm{UI}})b(s_i|s_{\mathrm{UI}})b(s_j|s_{\mathrm{UI}}). \tag{8.2}$$

To acquire exact goal states, a full Bayesian network must be established to track the set of overall states. Although we consider a slot-independent model for update to ignore dependencies between slots, its computational complexity could increase exponentially with the number of possible observed slots and system output types. We used a heuristic method that uses update rules to track individual states. The main advantage of the heuristic method is that its complexity is reasonable and that the method does not need prior dialog examples for training. Let H be a heuristic function built using update rules (Wang and Lemon 2013). Let the initial distribution of s be s_{init} and let the service action set be A_{service}. The update of belief probability $b \rightarrow b' : b(s'|s, a, o)$ can be represented as

$$b(s'|s, a, o) = \begin{cases} H(s, a, o) & (a \notin A_{\mathrm{service}}), \\ s_{\mathrm{init}} & (a \in A_{\mathrm{service}}). \end{cases} \tag{8.3}$$

8.3.2 Construction of Summary States

As the number of overall states could be large, learning a feasible POMDP policy may be a difficult task. In the proposed system, we introduced summary states that would be feature states in training POMDP policy. The motivation of using summary states is to prune out redundant information to reduce the number of states.

We used grid-based approximation (Thomson and Young 2010) to generate a summary state. The key idea of the grid approach is to use only information that is relevant to the two most-likely values, s_{x1} and s_{x2} for each slot x. The distribution between s_{x1} and s_{x2} can be approximated using grid points. Euclidean distance between grid points is measured and the nearest grid is selected for estimation. Appropriate selection of grid points is important. Previous work (Thomson and Young 2010) selected seven grid points for the distribution $l(s_{x1}, s_{x2})$: [(1.0, 0.0), (0.8, 0.2), (0.8, 0.0), (0.6, 0.4), (0.6, 0.2), (0.6, 0.0), (0.4, $*$)]. We selected 22 grid points to increase the accuracy of the approximation. We also considered two extra conditional features in constructing summary state for each slots: argument relation feature and none-value feature.

The argument-relation feature represents whether the given slot could be the appropriate argument with a given user intention. Suppose that there exists a slot "Actor name" of which values are the possible names of actors. "Actor name" can be an argument of the user intention "search-program," because users can utter names of actors to find a program. However, the slot cannot be an argument of the user intention, "volume-up," because users do not utter names when issuing this instruction.

The none-value feature represents whether the most-likely value of given slot is "none", the special value which represents that the slot is not mentioned by the user. The motivation is to represent the case that instances of the slot are not likely to be uttered by a user. In this case, the system should elicit a meta-action to request the value of slot, rather than ask users whether the value of slot is "none." Let the number of grid points be N_l.[1] The size of the feature vector assigned to the none-value feature is also N_l; this vector represents a bicomponent distribution l with the value s_{x1} = none.

The size of feature vector for each entity slot for one meta-action would be $N_c = 2N_l + 1$. Let $y_{i,l}$ be an indicator for corresponding grid points and let y_{none} and y_{arg} be indicators for the none-value feature and the argument-relation feature, respectively.

The kth value of $\phi(b_i)$ for the j-th meta action can be constructed as in Eq. (8.4). The construction is achieved by tiling corresponding features.

$$\phi_j(b_i)_k = \begin{cases} \begin{cases} y_{i,l} & \text{if } y_{\text{arg}} = 1 \text{ and } y_{\text{none}} = 0, \\ 0 & \text{otherwise,} \end{cases} & \text{if } 0 \le k < N_l, \\ \begin{cases} y_{i,l} & \text{if } y_{\text{arg}} = 1 \text{ and } y_{\text{none}} = 1 \\ 0 & \text{otherwise,} \end{cases} & \text{if } N_l \le k < 2N_l, \\ \begin{cases} 1 & \text{if } y_{\text{arg}} = 0, \\ 0 & \text{otherwise,} \end{cases} & \text{if } k = 2N_l. \end{cases} \tag{8.4}$$

8.4 Selection of Meta-Actions

The function of the meta-action selector is to choose which meta-action should be elicited for given tracked states. Meta-actions can be divided into two groups: *Submission* and *Confirmations*. When the "Submission" meta-action is invoked, the

[1]The notation is suggested by the previous research (Thomson and Young 2010).

Table 8.1 Interpretation of meta-action

Meta Action (a_{meta})	Interpretation
repeat	Requests the user to repeat
$request - x$	Requests the user to repeat the value of slot x
$confirm - x$	Confirms the user whether the value of slot x has one-best value
submit	Accepts slot values from the tracker

action selector accepts the slot values from the tracker and submits them to the service provider to generate a service system action. Otherwise, the action selector generates a corresponding meta-response and sends it to the user for confirmation. The "Confirmations" include "request user to repeat," "request user to utter the value of specified slot x" and "confirm user whether the value of slot x is y" (Table 8.1).

To acquire adequate meta-actions, POMDP policies must be optimized with given rewards. The policies can be described as a parameter vector θ in Gibbs sampling [Eq. (8.5)] and they can be calculated by any feasible reinforcement-learning technique. To acquired optimized POMDP policies, we used Episodic Natural Actor Critic algorithm (Peters and Schaal 2008; Thomson and Young 2010), which is a gradient descent method that uses gradients from the Fisher metric.

$$\pi(a|\bar{b}, \theta) = \frac{e^{\theta \cdot \phi_a(b)}}{\sum_{a'} e^{\theta \cdot \phi_{a'}(b)}}. \tag{8.5}$$

8.5 Generation of System Actions

Meta-actions elicited by the optimized policy must be interpreted before they are provided to end users, because meta-actions are internal messages for the system and do not include any information for users. Tracked result of belief states is accepted by the service provider (Sect. 8.4). The service provider selects a system action by using dialog model, which can be denoted as $P(a_m|o, h)$ where h is dialog history. When the model selects an appropriate action to be performed, the service provider uses a template database to generate a response sentence. The service provider also uses accepted dialog information to generate queries for content database to fill slot values (Fig. 8.2).

This separated architecture allows the system to provide various service actions. Previous architectures required that SDS designers build new a tracker model and that they assign new reward values when new system actions were added to a system. In contrast, the proposed approach requires only modification of the dialog model in the service provider. The main point of the suggested architecture is to bind valid system actions in an equivalent class in terms of performing service. This technique would be also effective for domains that have characteristics that are similar to those of the EPG domain.

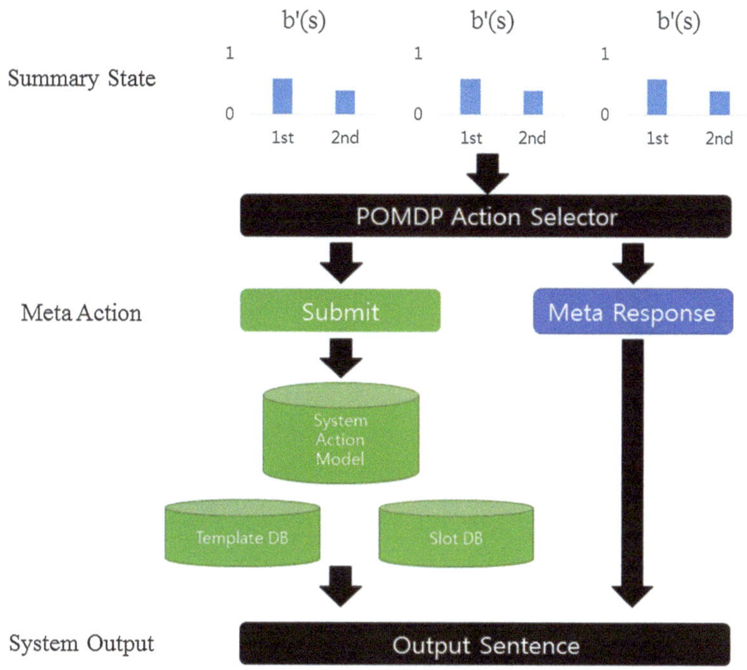

Fig. 8.2 Generation of system action using meta-action

8.6 Experiment

8.6.1 Setting

To verify the feasibility of proposed architecture, we iteratively trained POMDP model and observed its learning curve to assure its convergence. We measured average reward and turn length for each batch in the training process. Each training session started with policy parameters initialized to a zero vector. Each dialog batch consisted of 100 dialog instances. Two hundred batches were provided to the system and policy parameters were updated at the end of each batch. Each dialog instance consists of two, three, or four requests with corresponding constraints. At most 20 turns were permitted for each dialog instance.

The overall dialog was penalized for incorrect behaviors: requesting irrelevant slots, providing incorrect service, and wrongly confirming. Penalties differed for each case, but ranged from −5 to −40. A reward of +15 was given if the system gave correct responses. Otherwise, the reward was not given. The initial reward was set to zero for each dialog instance.

For training and evaluating, we used a simulator that is implemented by a request-constraint model (Schatzmann et al. 2007). Because the overall dialog session consists of multiple independent requests with drifting goals, we reconstructed the model of this simulator to make it appropriate for use in a multiple-goal SDS. In the original design, each of requests shares its constraint in a global constraint set. In our design, they are supported by isolated constraint sets, dividing request independently. ASR/NLU error rates were 0.1 for both user intention slots and entity slots.

8.6.2 Result

In the training sessions, the learning curves showed that the average reward and the average length of turn converged to certain level. Average length of turn converged to 4 (Fig. 8.3). Considering that the number of initial requests for each dialog instances ranged from 2 to 4, the converged level of average reward suggests that the proposed system can generate correct system actions for each user request with a manageable number of confirmation sentences. Average reward for each dialog instance converged to 0 (Fig. 8.4), which is slightly less than optimal reward 30–60. Since penalties for inappropriate confirmation (up to −40) were significantly larger than the reward value for a correct response, the converged reward lay in a predictable range. Reward is expected to increase if penalties are relieved.

Convergence of both learning curves indicates that the proposed architecture operates in a stabilized manner with appropriate decision-making process. It implies the feasibility of the overall suggested system, although further user-based experiment would be necessary to verify its utility in a direct manner.

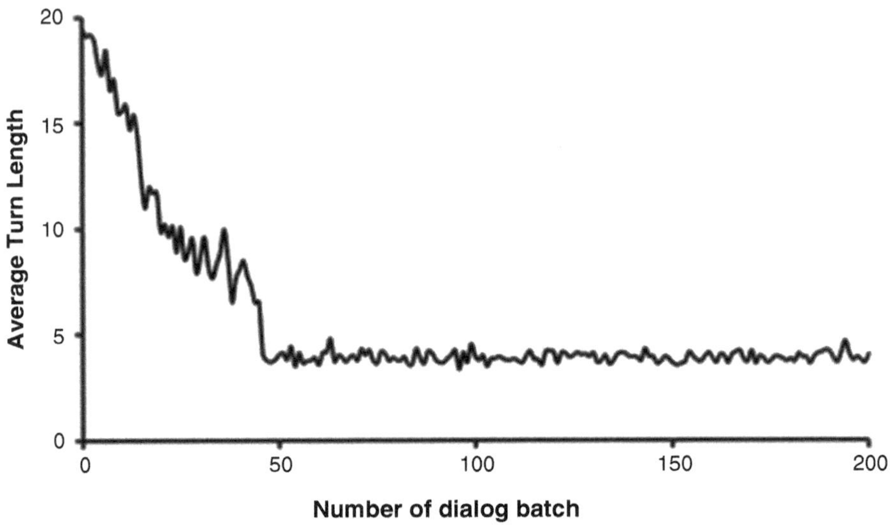

Fig. 8.3 Learning curves of average turn length

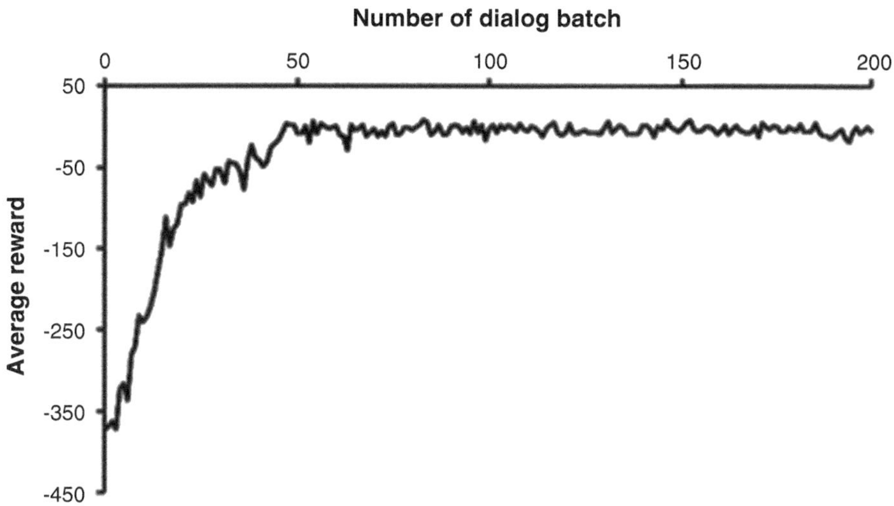

Fig. 8.4 Learning curves of average reward

8.7 Conclusion

This paper presents a new hybrid approach combined with the composite summary state POMDP components and stabilized service DM. The paper also introduces techniques to map summary states, which are used for POMDP action selection.

One further topic is a domain adaptation. Since the system was developed to overcome specific technical problems in the EPG domain, the proposed system is not guaranteed to operate effectively in other domains. Additional generalized components and models should be constructed in order to reflect characteristics of dialog acquired from other domains.

Another interesting topic would be to adopt iterative methods. By using a simulator and the proposed POMDP hybrid DM, log data of dialog could be generated. These log data could then be used to train the proposed DM and to construct an error model in simulator. We anticipate that iterative learning using real log data would boost the accuracy of overall dialog process. Further work will focus on methods to develop organized methods for these suggested iterative learning techniques.

Acknowledgements This work was supported by National Research Foundation of Korea (NRF) grant funded by the Korea government (MSIP) (NRF-2014R1A2A1A01003041).

References

Bellman R (1957) A Markovian decision process. Indiana Univ Math J 6:679–684

Bui T, Zwiers J, Poel M, Nijholt A (2006) Toward affective dialogue modeling using partially observable Markov decision processes. In: Reichardt D, Levi P, Meyer JJ (eds) Proceedings of the 1st workshop on emotion and computing? Current research and future impact. University of Bremen, Bremen, pp 47–50. http://www.doc.utwente.nl/66792/

Bui T, Poel M, Nijholt A, Zwiers J (2009) A tractable hybrid ddn-pomdp approach to affective dialogue modeling for probabilistic frame-based dialogue systems. Nat Lang Eng 15(2):273–307

Gasic M, Breslin C, Henderson M, Kim D, Szummer M, Thomson B, Tsiakoulis P, Young S (2013) Pomdp-based dialogue manager adaptation to extended domains. In: Proceedings of the SIGDIAL 2013 conference. Association for Computational Linguistics, Metz, pp 214–222. http://www.aclweb.org/anthology/W/W13/W13-4035

Peters J, Schaal S (2008) Natural actor-critic. Neurocomputing 71(7–9):1180–1190. doi:http://dx.doi.org/10.1016/j.neucom.2007.11.026. http://www.sciencedirect.com/science/article/pii/S0925231208000532. Progress in modeling, theory, and application of computational intelligence. 15th European symposium on artificial neural networks 2007. 15th European symposium on artificial neural networks 2007

Roy N, Pineau J, Thrun S (2000) Spoken dialogue management using probabilistic reasoning. In: Proceedings of the 38th annual meeting on association for computational linguistics, ACL '00. Association for Computational Linguistics, Stroudsburg, PA, pp 93–100. doi:10.3115/1075218.1075231. http://www.dx.doi.org/10.3115/1075218.1075231

Schatzmann J, Thomson B, Weilhammer K, Ye H, Young S (2007) Agenda-based user simulation for bootstrapping a pomdp dialogue system. In: Human language technologies 2007: the conference of the North American chapter of the association for computational linguistics; companion volume, short papers, NAACL-Short '07. Association for Computational Linguistics, Stroudsburg, PA, pp 149–152

Thomson B, Young S (2010) Bayesian update of dialogue state: a pomdp framework for spoken dialogue systems. Comput Speech Lang 24(4):562–588. doi:10.1016/j.csl.2009.07.003. http://www.dx.doi.org/10.1016/j.csl.2009.07.003

Wang Z, Lemon O (2013) A simple and generic belief tracking mechanism for the dialog state tracking challenge: on the believability of observed information. In: Proceedings of the SIGDIAL 2013 conference. Association for Computational Linguistics, Metz, pp 423–432. http://www.aclweb.org/anthology/W13-4067

Williams JD, Young S (2007) Partially observable Markov decision processes for spoken dialog systems. Comput Speech Lang 21(2):393–422

Young S (2006) Using pomdps for dialog management. In: Spoken language technology workshop, 2006. IEEE, pp 8–13. doi:10.1109/SLT.2006.326785

Young S, Schatzmann J, Weilhammer K, Ye H (2007) The hidden information state approach to dialog management. In: IEEE international conference on acoustics, speech and signal processing, 2007. ICASSP 2007, vol 4, pp. IV–149–IV–152. doi:10.1109/ICASSP.2007.367185

Young S, Gašić M, Keizer S, Mairesse F, Schatzmann J, Thomson B, Yu K (2010) The hidden information state model: a practical framework for pomdp-based spoken dialogue management. Comput Speech Lang 24(2):150–174. doi:10.1016/j.csl.2009.04.001. http://www.dx.doi.org/10.1016/j.csl.2009.04.001

Zhang B, Cai Q, Mao J, Guo B (2001) Planning and acting under uncertainty: a new model for spoken dialogue systems. In: Proceedings of the 17th conference on uncertainty in artificial intelligence, UAI'01. Morgan Kaufmann, San Francisco, CA, pp 572–579. http://www.dl.acm.org/citation.cfm?id=2074022.2074092

Chapter 9
A Voice QR Code for Mobile Devices

Donghyun Lee, Minkyu Lim, Minho Ryang, Kwang-Ho Kim, Gil-Jin Jang, Jeong-Sik Park, and Ji-Hwan Kim

Abstract This paper proposes a voice QR code for mobile devices. The QR code shows great performance for error correction and recovers decoding errors caused by skewed image angle or luminosity. In order to correct an image shot of the QR code symbol, a complex error code and data map need to be generated. Additionally, there is a need for an efficient QR code format and an audio codec for voice interface. This paper presents the generation method of the complex error code and data map in the voice QR code and suggests the efficient QR code format and an audio codec for voice interface.

Keywords Voice QR code • Voice interface • Audio codec • Error correction

9.1 Introduction

Voice interface has an important role in human–computer interaction environments (Edim and Muyingi 2014). For example, Intelligent Personal Assistants (IPAs) such as Siri and Google Now provide an eyes free environment to search and reserve restaurants using speech interface. This interface is already becoming popular and being used to the interaction between humans and mobile devices (Tsui et al. 2015; Walter et al. 2014).

D. Lee • M. Lim • M. Ryang • K.-H. Kim • J.-H. Kim (✉)
Department of Computer Science and Engineering, Sogang University,
Seoul 121-742, South Korea
e-mail: redizard@sogang.ac.kr; lmkhi@sogang.ac.kr; k2minho@sogang.ac.kr;
kimkwangho@sogang.ac.kr; kimjihwan@sogang.ac.kr

G.-J. Jang
School of Electronics Engineering, Kyungpook National University,
Daegu 702-701, South Korea
e-mail: gjang@knu.ac.kr

J.-S. Park
Department of Information and Communication Engineering, Yeungnam University,
Gyeongsan 712-749, South Korea
e-mail: parkjs@yu.ac.kr

G. Geunbae Lee et al. (eds.), *Natural Language Dialog Systems and Intelligent Assistants*, DOI 10.1007/978-3-319-19291-8_9

The Quick Response (QR) code system has gained tremendous popularity in recent years as mobiles have widely spread among the general public. This supplementary section presents a method to store word-level speech data in QR code symbols by using a mobile device; then an evaluation of the method follows (Soon 2008). A mobile device has the following advantages: (a) users incapable of typing can use the device easily, (b) there is no language barrier in using the device, and (c) the device is portable (Liu et al. 2008). For example, when the caregiver of a physically disabled person wants to create a QR code symbol containing the speech data signifying 'a medicine pill box', the user can simply press a button on the portable QR code printer, which is attached with a microphone. While the button is pressed, the microphone on the device is activated and speech is recorded until the button is released. After recording data, the speech data is compressed via an audio codec and stored in a QR code symbol to be printed on a label. The label with the QR code symbol printed on is attached on the medicine pill box by the user. As an example, the caregiver of a blind person can save the speech data of "ironed grey dress shirt" in a QR code symbol using a QR code printer. The blind person later opens a QR code decoder installed on the mobile. After taking a picture of the QR code symbol using the camera on the mobile, the QR code decoder generates corresponding speech data, "an ironed grey dress shirt" as recorded by the caregiver; the user perceives what the object is with the speech data 'an ironed grey dress shirt'.

The reason why we chose speech as the data to be stored in QR code symbols is that speech is the least restrictive medium of communication for humans. The QR code shows great performance for error correction and recovers decoding errors caused by skewed image angle or luminosity. The superb error recovery characteristic of QR code allows accurate data retrieval using stock apps and a mobile. This solution has price competitiveness when produced in small scales. This is an important factor to consider in designing a mobile device.

This paper proposes the implementation method of a voice QR code for speech interface. Our proposed method based on QR code has fast decoding speed and large capacity of data in high density using black-and-white two-dimensional barcode.

This paper is organized as follows. Section 2 introduces the related work in the QR code system. Section 3 describes the implementation method of the voice QR code. Section 4 concludes this paper.

9.2 Related Work

A QR code symbol is a black-and-white two-dimensional barcode. The system is characterized by its fast decoding speed and the large capacity of data in high density as compared to the conventional barcode system, and it also features error correction. The QR code symbol exploits vertical and horizontal dimensions while a barcode represents data only in one dimension, hence the QR code is capable of storing much larger amount of data. The QR code was initially applied to managing inventories, but soon began to replace barcodes and used on various printed media to facilitate Internet accesses with camera-attached mobile phones.

The conventional barcode stores around 20 alphanumeric characters, but the capacity of the QR code is larger by an order of magnitude. The QR code symbol can store up to 7089 numeric characters. As the QR code system uses both vertical and horizontal dimensions to store data, the amount of area needed to store a given data set is only one-tenth of the requirement of its corresponding barcode. The QR code system has a strong error correction capability and data can be recovered even for smudged code symbols. It can successfully recover error and decode data for QR codes with up to 30 % of the code area stained. A single set of data may be put in a series of QR codes. A data set may be divided by up to 16 symbols, and this allows printing codes on a thin and long area. Conversely, a data set stored in a series of QR code symbols may be merged onto a single QR code symbol.

9.3 Implementation Method of the Proposed Voice QR Code

In implementing voice QR code proposed in this paper, an issue has to be addressed: the amount of error correction code is greater than the amount of data itself. In order to correct an image shot of a QR code symbol, a complex error code and a data map need to be generated. Additionally, there is a need for an efficient voice QR code format and an audio codec. There is not a wide variety of stock hardware device or software program that meets our requirement to implement voice QR code, so it is not possible to use the result of a QR code decoder directly on an app.

The proposed voice QR code was implemented on a mobile device with a built-in microphone and a QR code printer. A software program is implemented that encodes input speech data with our proposed speech codec and renders it in the voice QR code format. A portable printer model is chosen that can print QR code symbols on labels among those available on the consumer market. Operating system and camera performance are the major considerations in deciding on a device, and a software program that allows for exploiting the decoding result of QR code symbol is selected.

In order to encode a one-second long speech data, the necessary data is 16 KB with the sampling rate of 8 K and 2B/sample. G.719 compresses speech data down to 1/16 of its original size, and one-second-long speech data under this codec requires 1 KB on the QR code. To implement the voice QR code, version 22 of the QR code with error correction level L is used. Under these options, that format can store up to 2 KB.

The QR code consists of rectangle-shaped symbol. A symbol consists of a quiet zone, a function pattern zone, and an encoded zone. The function pattern consists of position detection pattern, position detection divider, timing pattern, and alignment pattern. The encoded zone stores the format data, version information, and error correction code word. Function patterns are not used for data encoding.

9.4 Conclusions

This paper proposed a voice QR code for mobile devices. The QR code showed great performance for error correction and recovers decoding errors caused by skewed image angle or luminosity. This paper presented the generation method of the complex error code and data map in the voice QR code and suggested the efficient QR code format and an audio codec for voice interface.

Acknowledgements This research was supported by Basic Science Research Program through the National Research Foundation of Korea (NRF) funded by the Ministry of Science, ICT and Future Planning (No. NRF-2014R1A1A1002197).

References

Edim AE, Muyingi HN (2014) Speech user interface for low literacy users of ICT services. Eur J Comput Sci Inf Technol 2:18–29

Liu Y, Yang J, Liu M (2008) Recognition of QR code with mobile phones. Chinese Control Decis. Conf., pp 203–206

Soon TJ (2008) QR code. Synth J 3:59–78

Tsui KM, Dalphond JM, Brooks DJ, Medvedev MS, Mccann E, Allspaw J, Kontak D, Yanco HA (2015) Accessible human–robot interaction for telepresence robots: a case study. J Behav Robot 6:1–29. doi:10.1515/pjbr-2015-0001

Walter O, Despotovic V, Haeb-umbach R, Gemmeke JF, Ons B, Van H (2014) An evaluation of unsupervised acoustic model training for a dysarthric speech interface. Interspeech

Chapter 10
Detecting Multiple Domains from User's Utterance in Spoken Dialog System

Seonghan Ryu, Jaiyoun Song, Sangjun Koo, Soonchoul Kwon, and Gary Geunbae Lee

Abstract Multi-domain spoken dialog system should be able to detect more than one domain from a user's utterance. However, it is difficult to train an accurate binary classifier of a domain based on only positive and unlabeled examples. This paper improves hierarchical clustering algorithm to automatically identify reliable negative examples among unlabeled examples. This paper also verifies three linkage criteria that measure the distance between two clusters. In experiments, the proposed method resulted in the highest gain of F_1 score compared to the existing methods.

Keywords Dialog system • Domain selection • Domain detection • Learning from positive and unlabeled examples • Hierarchical clustering • Support vector machine

10.1 Introduction

Spoken dialog system (SDS) provides natural language interface between human and computer. Especially, multi-domain SDS (MDSDS) provides dialog service to many domains including *restaurant guide*, *car navigation*, *movie guide*, and *movie ticketing*. MDSDS first selects domain that the user may desire and then performs domain-specific processes: natural language understanding, dialog management, and response generation. Therefore, the selection of the appropriate domain from a user's utterance is a bottleneck of MDSDS; the incorrect selection of domain drives MDSDS to generate nonsense response.

The domain selection component usually uses as a multi-class classifier which is trained from multi-domain corpora. However, the boundaries of domains are ambiguous in the real world (Ryu et al. 2012). For example, when a user says "*I'm planning to go to Busan,*" the intended domain of the user could be *car navigation*, *hotel reservation*, or both. Therefore, for accurate service, the domain selection

S. Ryu (✉) • J. Song • S. Koo • S. Kwon • G.G. Lee
Pohang University of Science and Technology, Pohang, Republic of Korea
e-mail: ryush@postech.ac.kr; tichiel@postech.ac.kr; giantpanda@postech.ac.kr; theincluder@postech.ac.kr; gblee@postech.ac.kr

G. Geunbae Lee et al. (eds.), *Natural Language Dialog Systems and Intelligent Assistants*, DOI 10.1007/978-3-319-19291-8_10

component should be able to detect more than one domain at one time from a user's utterance. We called this task multi-domain detection (MDD).

MDD is a multi-label classification problem and this problem can be solved by combining in-domain verifiers (IDVs); an IDV classifies whether a user's utterance belongs to that domain. The IDV can be implemented as a binary classifier (BC) trained from positive examples and negative examples. Initially, a BC can be trained by considering the target domain's corpus as positive examples and the rest of the domain's corpora as negative examples. However, that BC can cause many incorrect rejections because the rest of the domain's corpora have some in-domain utterances. So the nature of the rest of the domain's corpora is *unlabeled* examples, not necessarily *negative* examples.

In this paper, we solved the MDD task by using a two-step approach to train an accurate BC from only positive and unlabeled examples. In the first step, we automatically identified reliable negative examples among unlabeled examples. We first constructed one cluster of positive examples and several clusters of unlabeled examples. Then we hierarchically merged clusters by fixing the cluster that was constructed from positive examples. We also verified three linkage criteria that measure the distance between two clusters.

In the second step, we trained a BC iteratively based on positive examples and the reliable negative examples identified in the first step. The obtained BC was more accurate than the basic BC that was trained using all the rest of the domain's corpora as negative examples. The effectiveness of this two-step approach in learning from positive and unlabeled examples has been demonstrated theoretically (Liu et al. 2002).

The remainder of this paper is organized as follows: Sect. 10.2 briefly introduces related work and explains the contributions of this paper compared to the related work. Section 10.3 describes the proposed method of hierarchical clustering in detail. Section 10.4 demonstrates the experimental design and results. Finally, Sect. 10.5 concludes the paper.

10.2 Related Work

Ryu et al. (2012) first introduced the MDD task and proposed an automatic multi-domain label annotation method that uses a hierarchical domain model designed by humans. The method automatically assigns positive and negative labels to utterances based on their previously annotated intents and named entities. The method performed well for a small-scale MDSDS. However, it is difficult for human to design complex hierarchical domain models. So the method can hardly be applied to large-scale MDSDS.

A similar task occurs in natural language question answering systems. In question answering systems, detecting multiple possible answer types for a question can give a chance to improve the answer. A two-step classification method can be used to solve this problem: the first step is to classify a question into several coarse-grained classes; the second step is to classify it into several fine-grained classes that belong to the coarse-grained classes (Li and Roth 2002). However, this method is based on classical multi-class classification and selects multiple answer types with an empirically determined confidence score threshold in the multi-class classification.

Some research considered learning from positive and unlabeled examples. PEBL (Yu et al. 2002) uses 1-disjunctive normal from (1-DNF) technique to identify reliable negative examples and then trains a support vector machine (SVM) iteratively. S-EM (Liu et al. 2002) first proposes a Spy technique to identify reliable negative examples and then uses Expectation Maximization (EM) (Dempster et al. 1997) to train a Naïve Bayes classifier (McCallum and Nigam 1998). Roc-SVM (Li and Liu 2003) uses an existing Rocchio method (Rocchio 1971) to identify reliable negative examples and then uses a classifier selection method to train an SVM iteratively. Biased-SVM (Liu et al. 2003) focused on biased formulation of SVM.

To our knowledge, no previous work applies learning from positive and unlabeled examples to MDD task. One contribution point of this paper is improving hierarchical clustering algorithm to identify reliable negative examples among unlabeled examples; we fixed a cluster which consists of positive examples at the beginning of hierarchical clustering. Another contribution point of this paper is verifying which linkage criterion works accurately for MDD task.

10.3 Methods

10.3.1 Hierarchical Clustering from Positive and Unlabeled Examples

10.3.1.1 Cluster Initialization

In our work, a cluster is a set of items and each item is a set of words in an example. We constructed clusters C from positive examples P and unlabeled examples U (Fig. 10.1; Algorithm 1 lines 1–2). We first constructed one cluster F of P and a total of $|U|$ clusters of each unlabeled example in U. So C initially consisted of $1 + |U|$ clusters.

10.3.1.2 Hierarchical Clustering

After initializing C, we performed the following hierarchical clustering iteratively (Algorithm 1 lines 3–12): merge two clusters X and Y when the distance $d_C(X, Y)$ between the two clusters is minimum in C. The distance measure is discussed in Sect. 10.3.2. When F and another cluster Z were selected to be merged during the iteration process, we did not perform the actual merge; we removed Z from C.

Algorithm 1 Hierarchical clustering from positive and unlabeled examples

Input
- $P = \{P_1, \ldots P_{|P|}\}$: positive examples
- $U = \{U_1, \ldots U_{|U|}\}$: unlabeled examples
- t_d: maximum cluster distance criterion ($0 < t_d \leq 1$)
- t_k: minimum cluster number criterion ($2 \leq t_k < 1 + |U|$)

Local variable
- $C = \{C_1, \ldots C_{|C|}\}$: clusters
- $RN = \{RN_1, \ldots RN_{|RN|}\}$: reliable negative examples
- F: a fixed cluster

Output
- Reliable negative examples

1. $F \leftarrow \{P_1, \ldots P_{|P|}\}$;
2. $C \leftarrow \{F, U_1, \ldots U_{|U|}\}$;
3. **while** $|C| > t_k$ **do**
4. $(i^*, j^*) \leftarrow (null, null)$; $minDist \leftarrow 1.0$;
5.   **foreach** (i, j), where $C_i \in C$, $C_j \in C$, and $C_i \neq C_j$ **do**
6.   **if** $d_C(C_i, C_j) < minDist$ **then**
7. $(i^*, j^*) \leftarrow (i, j)$; $minDist \leftarrow d_C(C_i, C_j)$;
8.   **if** $minDist < t_d$ **then**
9.   **if** $C_{i^*} = F$ **then**$C \leftarrow C - C_{j^*}$;
10.   **else if** $C_{j^*} = F$ **then** $C \leftarrow C - C_{i^*}$;
11.   **else** $C \leftarrow C - C_{i^*} - C_{j^*}$; $C \leftarrow C \cup \{C_{i^*} \cup C_{j^*}\}$;
12.   **else** escape loop;
13. **for** $i = 1$ to $|C|$, **where** $C_i \neq F$ **do**
14.   **for** $k = 1$ to $|C_i|$ **do**
15. $RN \leftarrow RN \cup \{C_{i,k}\}$;
16. **return** RN;

We did this because we focused in only identifying reliable negative examples and were not interested in identifying additional positive examples. We terminated hierarchical clustering when the minimum distance between two clusters in C was too far or $|C|$ was sufficiently small.

10.3.1.3 Reliable Negative Examples Selection

After hierarchical clustering, we regarded the remaining clusters in C as reliable negative examples RN except for F (Algorithm 1 lines 13–15). For example (Fig. 10.1), when $P = \{p_1, \ldots p_3\}$ and $U = \{u_1, \ldots u_6\}$, a total of seven clusters

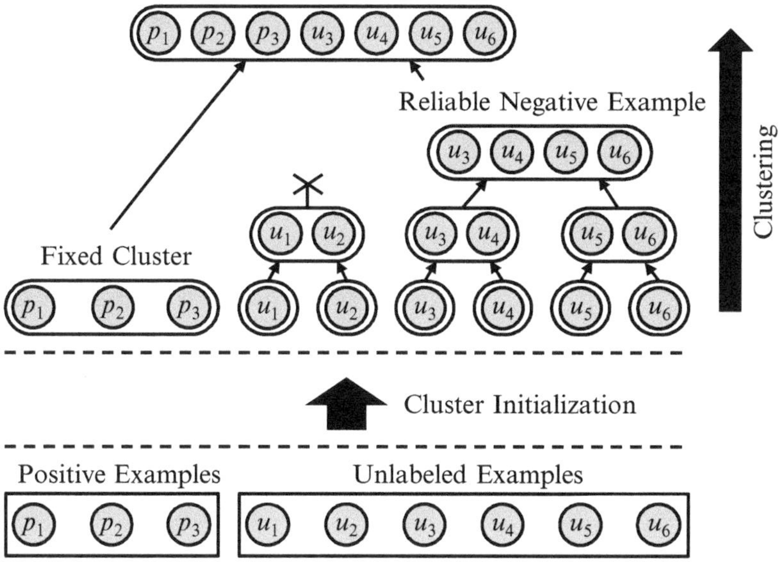

Fig. 10.1 An example of the proposed hierarchical clustering from positive and unlabeled examples: u_3, u_4, and u_5 are reliable negative examples

are constructed: $F = \{p_1, \ldots p_3\}$ from P and six clusters $\{u_1\}, \ldots \{u_6\}$ from each unlabeled example in U. When distance between $\{u_1, u_2\}$ and F reached a minimum value during the iteration process, $\{u_1, u_2\}$ was removed from C. When F and $\{u_3, \ldots u_6\}$ remain at the end of iteration, $\{u_3, \ldots u_6\}$ is selected as *RN* examples.

10.3.2 Linkage Criteria

We defined an item as a set of words. The distance $d_J(x, y)$ between two items x and y is the Jaccard distance:

$$d_J(x, y) = 1 - J(x, y) = 1 - \frac{|x \cap y|}{|x \cup y|}, \qquad (10.1)$$

where $J(x, y)$ is the Jaccard similarity between x and y. So $|x \cap y|$ is the number of words in the intersection set of x and y; $|x \cup y|$ is the number of words in the union set of x and y. This Jaccard distance is used in all linkage criteria below.

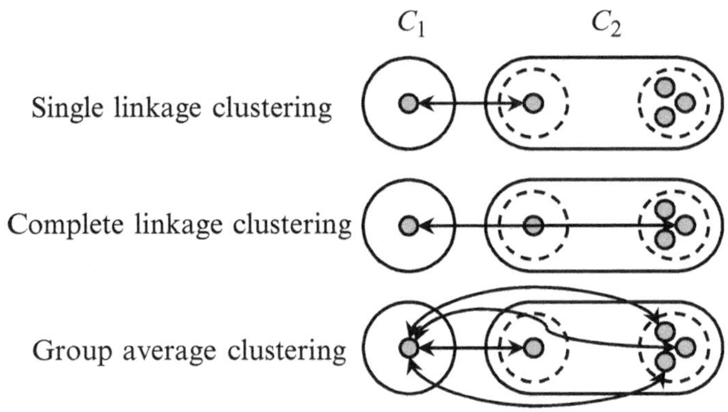

Fig. 10.2 Examples of the linkage criteria for MDD (*filled circles*: items, *solid-line empty circles*: clusters, *dashed-line empty circles*: implicit sub-clusters, *arrows*: considered distances in linkage criteria)

In hierarchical clustering, the selected clusters to be merged are the closest pair based on one of the following linkage criteria: single linkage clustering $d_{SL}(X, Y)$, complete linkage clustering $d_{CL}(X, Y)$, or group average clustering $d_{GA}(X, Y)$ (Hastie et al. 2009).

$$d_{SL}(X, Y) = \min_{x \in X,\, y \in Y} d_J(x, y) \tag{10.2}$$

$$d_{CL}(X, Y) = \max_{x \in X,\, y \in Y} d_J(x, y) \tag{10.3}$$

$$d_{GA}(X, Y) = \frac{1}{|X|\,|Y|} \sum_{x \in X} \sum_{y \in Y} d_J(x, y) \tag{10.4}$$

However, some methods do not work well in MDD, because a domain in MDSDS includes various types of utterances. For example, both two utterances "*City Cinema in Gangnam.*" and "*A Werewolf Boy seems interesting*!" are very different but both can be located in F for the *Movie Ticketing* domain. Therefore, a cluster can contain implicit sub-clusters: this structure can cause complete linkage clustering (10.3) and group average clustering (10.4) to fail.

For example, C_1 is close to the left sub-cluster of C_2 (Fig. 10.2), where C_1 is constructed from an unlabeled example and C_2 is constructed from positive examples. Therefore, C_1 and C_2 should be merged. Single linkage clustering (10.2) gives a short distance between C_1 and C_2 so they are merged. In contrast, complete linkage clustering (10.3) and group average clustering (10.4) give a large distance between C_1 and C_2 so they are not merged. Therefore, we expected that single linkage clustering is more accurate than the other linkage criteria.

Algorithm 2 Iterative binary classifier training

Input

- $P = \{P_1, \ldots P_{|P|}\}$: positive examples
- $RU = \{RU_1, \ldots RU_{|RU|}\}$: remaining unlabeled examples
- $RN = \{RN_1, \ldots RN_{|RN|}\}$: reliable negative examples

Local variable

- Ω: binary classifier

Output

- Final binary classifier

loop
    Train Ω using P and RN;
    $N_{out} \leftarrow null$;
    for $i = 1$ to $|RU|$ **do**
        $c \leftarrow$ classify RU_i using Ω;
        if c is negative **then** $N_{out} \leftarrow N_{out} \cup \{RU_i\}$;
    if $|N_{out}| > 0$ **then** $RN \leftarrow RN \cup N_{out}$; $RU \leftarrow RU - N_{out}$;
    else escape loop
return Ω;

10.3.3 Iterative Training of the Binary Classifier

We obtained the final BC by training BCs iteratively using positive examples *P*, reliable negative examples *RN*, and the remaining unlabeled examples *RU* (Algorithm 2). We first used *P* and *RN* to train a BC Ω. Then we classified *RU* and obtained a set of negative outputs N_{out}. We added N_{out} to *RN* and removed N_{out} from *RU*. We repeated this iteration until Ω converged. We used LIBSVM (Chang and Lin 2011) as a BC: we used the radial basis function kernel, disabled shrinking heuristics, and used default settings for the remaining parameters in the experiment.

10.4 Experiments

10.4.1 Experimental Designs

We prepared Korean corpora of seven domains (Table 10.1). We used 80 % of the corpora as training data and 20 % of the corpora as test data. Each BC used the target domain's corpus as positive examples and the other six domains' corpora as unlabeled examples. For evaluation we labeled the test data as positive or negative.

We performed experiments with three existing methods and the proposed method.

Table 10.1 The basic information of collected corpora

Domain	Translated sentence example
D_1: Car Navigation	"*Please guide me the best path from Pohang to Gyeongju.*"
D_2: Civil Application Service	"*I want to renew my passport.*"
D_3: Home Control	"*What is in my refrigerator?*"
D_4: Movie Ticketing	"*City Cinema in Gangnam.*"
D_5: Traffic Guide	"*I'm planning to go to Busan.*"
D_6: Travel Reservation	"*I'm going to take a trip from Seoul to Busan.*"
D_7: Weather Information	"*How will the weather be on Sunday?*"

- Baseline: we trained the SVM of a domain by using the rest of the six domain's corpora as negative examples directly.
- OC-SVM: we trained the one-class SVM (OC-SVM) (Schölkopf et al. 2001) of a domain by using the target domain's corpus as positive examples and no negative examples.
- PEBL: we trained the SVM of a domain based on the PEBL framework (Yu et al. 2002).
- HCPU: we trained the SVM of a domain based on our hierarchical clustering from positive and unlabeled examples (HCPU). In HCPU, we tried three different linkage criteria: single linkage clustering (HCPU-SL), complete linkage clustering (HCPU-CL), and group average clustering (HCPU-GA).

We evaluated MDD performance by measuring the precision, recall, and F_1 score of each domain's BC. We also computed the macro-average precision, recall, and F_1 score.

10.4.2 Experimental Results

The proposed method HCPU-SL resulted in the highest gain in F_1 scores from baseline (Table 10.2): The macro-average F_1 score increased from 0.6070 (baseline) to 0.8110 (HCPU-SL), because HCPU increased macro-average recall from 0.4592 to 0.7352 without decreasing macro-average precision. In contrast, HCPU-CL and HCPU-GA had no significant change compared to baseline. Both OC-SVM and PEBL increased F_1 scores by increasing recall but they decreased precision.

10.5 Conclusion

We improved a method of hierarchical clustering from positive and unlabeled examples to solve the MDD task. In the experimental results, the proposed method had higher F_1 score than the existing methods (Fig. 10.3). The proposed method

Table 10.2 The precision, recall, and F_1 scores of MDD

	Baseline	OC-SVM	PEBL		HCPU	
				SL	CL	GA
(a) Precision						
D_1	0.9535	0.7334	0.9286	0.9322	**0.9537**	0.9514
D_2	**0.9370**	0.7547	0.9132	0.9189	0.9353	0.9344
D_3	0.8631	0.7188	0.8503	**0.8681**	0.8629	0.8611
D_4	0.9138	0.7701	0.8824	0.9111	**0.9175**	0.9122
D_5	0.9085	0.7388	0.8487	0.9031	**0.9126**	0.9100
D_6	0.8561	0.7866	0.8555	**0.9117**	0.8539	0.5814
D_7	**0.9285**	0.7010	0.8701	0.9159	0.9274	0.9274
Avg.	0.9086	0.7433	0.8784	0.9087	**0.9090**	0.8683
(b) Recall						
D_1	0.5115	**0.7430**	0.5845	0.6949	0.5140	0.5246
D_2	0.5328	**0.7803**	0.6528	0.7032	0.4906	0.4924
D_3	0.4267	0.8185	0.6162	**0.8419**	0.4173	0.4193
D_4	0.3640	0.7756	0.4283	**0.8116**	0.3573	0.3568
D_5	0.5602	**0.7562**	0.5639	0.6998	0.5633	0.5583
D_6	0.3412	**0.7458**	0.5511	0.7071	0.3412	0.3367
D_7	0.4779	**0.7724**	0.6368	0.6877	0.4859	0.4859
Avg.	0.4592	**0.7703**	0.5762	0.7352	0.45288	0.4534
(c) F_1 score						
D_1	0.6659	0.7381	0.7175	**0.7963**	0.6680	0.6732
D_2	0.6793	0.7673	0.7614	**0.7967**	0.6436	0.6449
D_3	0.5710	0.7654	0.7146	**0.8548**	0.5625	0.5640
D_4	0.5206	0.7728	0.5767	**0.8585**	0.5144	0.5129
D_5	0.6930	0.7474	0.6776	**0.7885**	0.6966	0.6920
D_6	0.4879	0.7657	0.6703	**0.7965**	0.4876	0.4829
D_7	0.6310	0.7350	0.7354	**0.7856**	0.6377	0.6377
Avg.	0.6070	0.7560	0.6933	**0.8110**	0.6015	0.6011

reduced the number of false-negative errors and therefore achieved high recall compared to the baseline (Fig. 10.3). This is because the final BC was trained iteratively using identified reliable negative examples. We also verified that single linkage clustering is the most accurate linkage criterion for the MDD task. This is because the other linkage criteria identified incorrectly most unlabeled examples as negative examples.

We plan to perform research on out-of-domain (OOD) detection. MDSDS should detect OOD utterances and reject them. The problem is that detecting OOD without using actual OOD data for training is a difficult task (Lane et al. 2007). However, we expect OOD detection problem can be solved by applying the proposed method into large-scale unlabeled examples such as conversational logs.

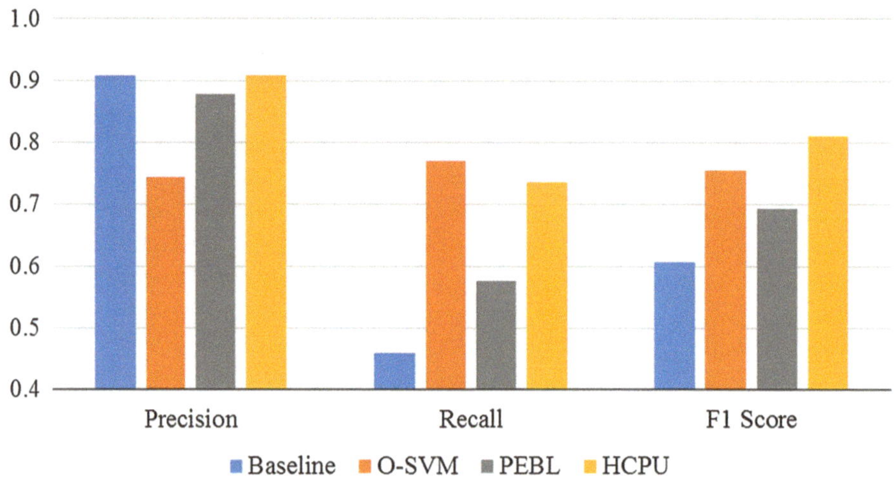

Fig. 10.3 Summary of MDD experiments

Acknowledgments This work was supported by ICT R&D program of MSIP/IITP [14-824-09-014, Basic Software Research in Human-level Lifelong Machine Learning (Machine Learning Center)]. This work was supported by National Research Foundation of Korean (NRF) [NRF-2014R1A2A1A01003041, Development of Multi-party Anticipatory Knowledge-Intensive Natural Language Dialog System].

References

Chang C, Lin C (2011) LIBSVM: a library for support vector machines. ACM Trans Intell Syst Technol 2(3):27:1–27:27

Dempster AP, Laird NM, Rubin DB (1997) Maximum likelihood from incomplete data via the EM algorithm. J R Stat Soc Ser B Methodol 39(1):1–38

Hastie T, Tibshirani R, Friedman J (2009) The elements of statistical learning, 2nd edn. Springer, New York, pp 520–528

Lane I, Kawahara T, Matsui T, Nakamura S (2007) Out-of-domain utterance detection using classification confidences of multiple topics. IEEE Trans Audio Speech Lang Process 15(1):150–161

Li X, Liu B (2003) Learning to classify texts using positive and unlabeled data. In: Proceedings of the 18th international joint conference on artificial intelligence, Acapulco, Mexico, August 2003

Li X, Roth D (2002) Learning question classifiers. In: Proceedings of the 19th international conference on computational linguistics, Taipei, Taiwan, September 2002

Liu B, Lee WS, Yu PS, Li X (2002) Partially supervised classification of text documents. In: Proceedings of the 19th international conference on machine learning, New South Wales, Sydney, July 2002

Liu B, Dai Y, Li X, Lee WS, Yu PS (2003) Building text classifiers using positive and unlabeled examples. In: Proceedings of the 3rd IEEE international conference on data mining, Melbourne, Florida, USA, November 2003

McCallum A, Nigam K (1998) A comparison of event models for Naive Bayes text classification. In: Proceedings of the 15th natural conference on artificial intelligence: workshop on learning from text categorization, Madison, Wisconsin, USA, July 1998

Rocchio J (1971) Relevance feedback in information retrieval. In: The smart retrieval system: experiments in automatic document processing, Englewood Cliffs, New Jersey, USA, 1971

Ryu S, Lee D, Lee I, Han S, Lee GG, Kim M, Kim K (2012) A hierarchical domain model-based multi-domain selection framework for multi-domain dialog systems. In: Proceedings of the 24th international conference on computational linguistics, Mumbai, India, December 2012

Schölkopf B, Platt JC, Shawe-Taylor J, Smola AJ, Williamson RC (2001) Estimating the support of a high-dimensional distribution. Neural Comput 13(7):1443–1471

Yu H, Han J, Chang KC (2002) PEBL: positive example based learning for web page classification using SVM. In: Proceedings of the 8th ACM SIGKDD international conference of knowledge discovery and data mining, Edmonton, Alberta, Canada, July 2002

Chapter 11
DietTalk: Diet and Health Assistant Based on Spoken Dialog System

Sohyeon Jung, Seonghan Ryu, Sangdo Han, and Gary Geunbae Lee

Abstract This paper presents DietTalk, a diet and health assistant based on a spoken dialog system. The purpose of DietTalk is to help people to control their weight by consulting with it using natural language. DietTalk stores personal status, provides food and exercise information, and recommends appropriate food and exercise. To evaluate the effectiveness of DietTalk, we performed human user experiments. DietTalk had good accuracy and satisfied users; therefore, DietTalk is effective in helping users to control their weight.

Keywords Spoken dialog system • Recommendation • Health care

11.1 Introduction

Many people intake an unbalanced diet or do not get sufficient exercise, so they have difficulty in achieving diet and health goals. Computer-assisted food and exercise management can help those people. A previous study proposed application software for diet and health management (Lim et al. 2011), but provided only a touch user interface; a dialog interface can provide better accessibility and user experience for users than can a touch interface. Due to recent advances in natural language processing technology, spoken dialog systems (SDSs) are now widely used to provide dialog interfaces (Lee et al. 2009, 2010).

In this paper, we propose DietTalk, which is a diet and health assistant based on SDS. DietTalk consists of a dialog agent (DA) and a service agent (SA) (Fig. 11.1). The DA understands requests, manages dialog, and generates responses. Based on the decision of the DA, the SA performs service functions including management of

S. Jung • S. Ryu (✉) • S. Han • G.G. Lee
Pohang University of Science and Technology, Pohang, Republic of Korea
e-mail: thankyoualot@postech.ac.kr; ryush@postech.ac.kr; hansd@postech.ac.kr; gblee@postech.ac.kr

G. Geunbae Lee et al. (eds.), *Natural Language Dialog Systems and Intelligent Assistants*, DOI 10.1007/978-3-319-19291-8_11

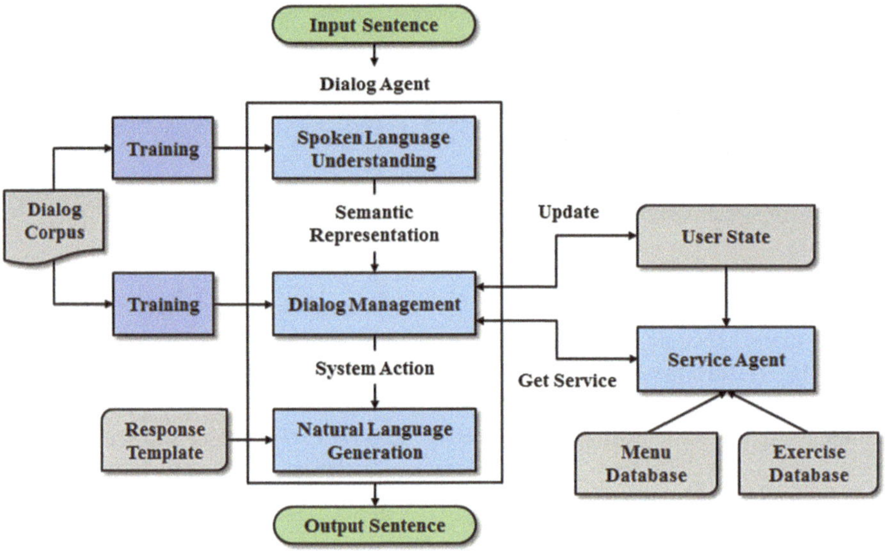

Fig. 11.1 System architecture of diet and health assistant based on spoken dialog system

personal status, offering of information on health and diet, recommendation of food menu, and recommendation of exercises. We proposed the concept of DietTalk in The REAL Challenge Workshop.[1]

The rest of the paper is organized as follows: Sect. 11.2 describes the DA in detail. Section 11.3 describes the SA in detail. Section 11.4 demonstrates the evaluation design and results. Finally, Sect. 11.5 draws conclusions.

11.2 Dialog Agent

DA consists of three sequential processes: spoken language understanding (SLU), dialog management (DM), and natural language generation (NLG). In SLU, a machine-readable semantic frame is automatically extracted from a natural language sentence. We defined a semantic frame to be a combination of intent and named entities. To determine intent we used maximum entropy (Ratnaparkhi and Marcus 1998) and to recognize named entities we used conditional random fields (Lafferty et al. 2001).

In DM, a system action is automatically chosen based on the dialog state and the extracted semantic frame. To determine a system action, we used the example-based dialog management (EBDM) technique (Lee et al. 2009). In EBDM, given

[1]https://dialrc.org/realchallenge/

the dialog state and the extracted semantic frame, the semantically closest example is retrieved from a dialog example database. Afterwards, the system action attached to the dialog example is chosen as the output of the DM.

In NLG, a natural language response is automatically generated based on the chosen system action. To generate a natural language response, we used a simple template-based approach (Lee et al. 2009). A sentence template is first selected from the system action, then the blanks are filled in with the parameters of the chosen action.

11.3 Service Agent

The SA performs a designated service and returns the service execution results. We implemented many service functions related to health and diet:

1) storing personal physical status (age, sex, height, and weight),
2) storing personal preferences on food and exercise,
3) storing desired weight,
4) offering information on health and diet,
5) recommending food menu,
6) recommending exercise.

In the rest of this section, we introduce the methods for recommending food menu and exercise.

11.3.1 Food Menu Recommendation

The SA uses four scores on a scale of 0 to 100 to evaluate a food menu M.

1. $\text{Score}_{\text{calory}}(M)$: closeness between total calories of M and adequate calories for the user's meal. A high score means that M provides appropriate calories.
2. $\text{Score}_{\text{pref}}(M, U)$: agreement between M and the stored preferences of the U. A high score means that M meets the U's preferences.
3. $\text{Score}_{\text{comb}}(M)$: the naturalness of M. We compute a combination score based on food category. A high score means that the combination of food items is natural for a human.
4. $\text{Score}_{\text{rand}}$: random score. This score allows users to receive various food recommendations even in the same situation.

The total score is the weighted sum of scores 1–4. To broaden user's choices, we designed three food menus. The SA recommends the food menu that has the maximum score in each type.

A type-1-menu consists of a single meal. The total score of this menu is computed as

$$0.75 \times \text{Score}_{\text{calory}}(M) + 0.15 \times \text{Score}_{\text{pref}}(M) + 0.1 \times \text{Score}_{\text{rand}} \tag{11.1}$$

A type-2-menu consists of a meal and a drink; a type-3-menu is a Korean meal that consists of rice, soup, and two side dishes. Total scores of type-2- and type-3-menus are computed as

$$0.65\times\text{Score}_{\text{calory}}(M)+0.15\times\text{Score}_{\text{pref}}(M)+0.1\times\text{Score}_{\text{comb}}(M)+0.1\times\text{Score}_{\text{rand}} \tag{11.2}$$

The food menu database was modified from a previous study (Lim et al. 2011).

11.3.2 Exercise Recommendation

To maximize the efficiency of calorie expenditure, the SA recommends Exercise (E) as combination of anaerobic and aerobic exercise.

1. $\text{Calorie}_{\text{today\_exercise}}(E)$: the calories which remain to be burned by exercise today. It considers $\text{Calorie}_{\text{remain}}(E)$ and $\text{Calorie}_{\text{diet}}(E)$. A high value means that the user has to exercise for much of the day.
2. $\text{Calorie}_{\text{remain}}(E)$: the difference between today's suggested calories and food calories which the user has eaten to the time of assessment. A high value means that the user did not eat many calories. A negative number means that the user has already exceeded today's suggested calorie intake.
3. $\text{Calorie}_{\text{diet}}(E)$: the number of calories to be burned to achieve the desired weight during the period set by the user. This variable considers current weight, goal weight, and the time (months) that the user has chosen as the period over which to lose the target amount of weight. A high value means that the user wants to lose a large amount of weight in a short time.

To lose 1 kg, the user must burn 7700 kcal more than he or she eats. Therefore, if current weight exceeds target weight, then based on $\text{Calorie}_{\text{today\_exercise}}(E)$, DietTalk chooses appropriate exercises from an exercise database. Calories that the user will burn by exercise are calculated as

$$\text{Calory}_{\text{diet}}(E) = \frac{(\text{Current weight} - \text{Goal weight}) \times 7700\text{ kcal}}{\text{User aimed days}} \tag{11.3}$$

$$\text{Calory}_{\text{today\_exercise}}(E) = \text{Calory}_{\text{diet}}(E) - \text{Calory}_{\text{remain}}(E) \tag{11.4}$$

We built the exercise database from various web pages about exercise.

11.4 Evaluation

11.4.1 Experimental Designs

To train SLU and the DM, we collected a Korean-language dialog corpus for the DietTalk domain. The corpus consists of 559 pairs of user utterance and system utterance. We annotated semantic information including intents ($n = 49$) and named entities ($n = 17$) onto each sentence.

To evaluate DietTalk, we performed human user experiments. We asked five student volunteers to complete five dialog tasks, then measured task completion rate (TCR), successful turn rate (STR), and average dialog length (ADL). We also asked the volunteers to complete a questionnaire.

11.4.2 Experimental Results

DietTalk achieved TCR = 0.84, STR = 0.81, and ADL = 8.80 (Table 11.1). This result indicates that DietTalk provided an accurate dialog interface to the volunteers.

Before using DietTalk, all volunteers had high interest in weight control, but were only moderately (5.6/10) satisfied with their weight control (Table 11.2). After using DietTalk, some volunteers pointed out that the naturalness of dialog was insufficient (Table 11.3). However, all volunteers were satisfied with the effectiveness of DietTalk for weight control (Table 11.3).

Table 11.1 Evaluated performance of DietTalk

Task completion rate	Successful turn rate	Average dialog length
0.84	0.81	8.80

Table 11.2 Users' questionnaires for DietTalk on a scale of 1–10

Questions	Mean	s.d.
Pre-test		
Do you have interest in weight control?	7.60	1.02
Do you fulfill your weight control compared to your interest?	5.60	1.62
Post-test		
Was your conversion with DietTalk natural?	7.20	2.14
Are you satisfied with response time of DietTalk?	7.80	1.47
Do you think DietTalk is useful for weight control?	8.40	1.36
Will you recommend DietTalk to friends?	8.00	1.10

Table 11.3 A translated example of dialog with DietTalk

Beginning of dialog	
User	My height is 175 cm and weight is 65 kg.
System	Yes. I stored this information (height: 175 cm, weight: 65 kg).
User	What is my BMI index?
System	Your BMI index is 21.22, and it is normal.
At lunch time	
User	How much calories are there in a slice of pizza?
System	A slice of pizza is 403.5 kcal.
User	Then I will eat whopper burger in lunch.
System	You ate whopper burger. OK.
At dinner time	
User	What shall I have for dinner?
System	How about a bowl of Bibimbap and a cup of Sikhye?
After having dinner	
User	Recommend exercise, please.
System	Today, what about do sideband dumb-bell and basketball 12 min?
User	Exercise is too tiring.
System	Cheer up!

11.5 Conclusion

In this paper, we proposed DietTalk, which is a diet and health assistant based on SDS technology. DietTalk consists of a DA and an SA. The dialog understands requests, manages dialog, and generates responses. The SA performs various service functions. DietTalk can provide personalized diet and health assistance service to users. In experiments, DietTalk had good accuracy and satisfied users; therefore, DietTalk may be effective in helping users to control their weight.

Acknowledgments This work was supported by National Research Foundation of Korean (NRF) [NRF-2014R1A2A1A01003041, Development of multi-party anticipatory knowledge-intensive natural language dialog system].

References

Lafferty JD, McCallum A, Pereira FCN (2001) Conditional random fields: probabilistic models for segmenting and labeling sequence data. In: Proc. of ICML 2001

Lee C, Jung S, Kim S, Lee GG (2009) Example-based dialog modeling for practical multi-domain dialog system. Speech Commun 15(5):466–484

Lee C, Jung S, Kim K, Lee D, Lee GG (2010) Recent approaches to dialog management for spoken dialog systems. J Comput Sci Eng 4(1):1–22

Lim BK, Kim JS, Yoo JH, Zhang BT (2011) DietAdvisor: a personalized eHealth agent in mobile computing environment. J KIISE 38(2D):115–118

Ratnaparkhi A, Marcus MP (1998) Maximum entropy models for natural language ambiguity resolution. Ph.D. Thesis, UPenn

Chapter 12
Lexicon Optimization for WFST-Based Speech Recognition Using Acoustic Distance Based Confusability Measure and G2P Conversion

Nam Kyun Kim, Woo Kyeong Seong, and Hong Kook Kim

Abstract In this paper, we propose a lexicon optimization method based on a confusability measure (CM) to develop a large vocabulary continuous speech recognition (LVCSR) system with unseen words. When a lexicon is built or expanded for unseen words by using grapheme-to-phoneme (G2P) conversion, the lexicon size increases because G2P is generally realized by 1-to-N-best mapping. Thus, the proposed method attempts to prune the confusable words in the lexicon by a CM defined as the acoustic model distance between two phonemic sequences. It is demonstrated through the LVCSR experiments that the proposed lexicon optimization method achieves a relative word error rate (WER) reduction of 14.72 % in a *Wall Street Journal* task compared to the 1-to-4-best G2P converted lexicon approach.

Keywords Lexicon optimization • Confusability measure • Grapheme-to-phoneme conversion • Weighted finite-state transducer

12.1 Introduction

Recently, many research works have been proposed in order to develop large vocabulary continuous speech recognition (LVCSR) systems, such as feature extraction, acoustic modeling, pronunciation modeling, language modeling, and decoding (Saon and Chien 2012). Among them, decoding or searching for word sequences with acoustic feature vectors plays a main role in the performance of LVCSR systems in which decision-tree-based or weighted finite-state transducer (WFST) approaches have been typically used for LVCSR decoding (Kanthak et al. 2002). The decision-tree-based approach requires a small amount of decoding memory. However, since on-the-fly composition must be performed with language models (LMs) during the recognition of test utterances, this approach increases

N.K. Kim • W.K. Seong • H.K. Kim (✉)
Gwangju Institute of Science and Technology (GIST), Gwangju, Republic of Korea
e-mail: skarbs001@gist.ac.kr; wkseong@gist.ac.kr; hongkook@gist.ac.kr

G. Geunbae Lee et al. (eds.), *Natural Language Dialog Systems and Intelligent Assistants*, DOI 10.1007/978-3-319-19291-8_12

decoding time (Kanthak et al. 2002). Conversely, a WFST for LVCSR decoding can generally be constructed by the composition of different speech recognition knowledge sources, such as a hidden Markov model (HMM) topology, a context-dependent phone model, a lexicon, and an n-gram LM, where each source is also represented by an individual WFST (Mohri et al. 2008). Thus, due to such modular representation and optimization techniques, a WFST-based decoder can offer simpler realization and faster decoding than a decision-tree-based decoder (Mohri et al. 2008).

When the domain for an LVCSR system is dynamically changed due to new-coined words, a word lexicon must be reconstructed to accommodate unseen words by using a data-driven approach. A method for dealing with this problem is grapheme-to-phoneme (G2P) conversion of such unseen words, which can be used for an expanded lexicon (Bisani and Ney 2008). However, the accuracy of G2P conversion depends on how much knowledge is incorporated into the design of the G2P conversion (Bisani and Ney 2008). Thus, it is more effective to make multiple pronunciations for a given unseen word by using an N-best G2P conversion, which unfortunately results in an excessive increase of the lexicon size and a further increase of the LVCSR decoder size. Consequently, this causes to increase the word error rate (WER) of the LVCSR system (Vinyals et al. 2009).

In order to prevent from increasing the WER while reducing the decoder size, eliminating the unnecessary nodes of a decision tree was proposed for a decision-tree-based decoder (Neukirchen et al. 1997). This approach achieved the reduced size of the decoder, but the WER of the reduced decision-tree-based decoder was similar to that of the original decision-tree-based one. For WFST-based decoders, several structural optimization techniques were proposed by sharing silence and short-pause states and restructuring the beam according to the token path (Guo et al. 2012). While this approach efficiently optimized the WFST, it was difficult to apply to the unseen word problem. In addition, a minimum classification error (MCE) model (Lin and Yvon 2007) and a conditional random field (CRF) model (Kubo et al. 2012) were proposed to optimize the decoding network size during WFST training. However, these methods needed to be applied repeatedly to retrain the WFST if unseen words were given. As an alternative, the decoding network size was reduced by using a confusability measure (CM) (Kim et al. 2008). This approach reduced the size, but it suffered from the excessive removal of words, causing an out-of-vocabulary problem (Jitsuhiro et al. 1998).

In this paper, we propose a method to optimize a G2P-converted lexicon that is realized by the N-best phoneme sequences of each word. To this end, a CM is first defined by the acoustic distance between two phoneme sequences and the length of the phoneme sequences. A G2P model-based N-best lexicon is then constructed to find the most probable phoneme sequence of an unseen word. However, since the lexicon becomes oversized, the lexicon is then optimized by pruning the confusable phoneme sequences using the CM.

Following this introduction, Sect. 12.2 briefly explains a lexicon construction using a G2P model. Section 12.3 describes the CM using the acoustic models and the dynamic programming-based alignment between the two phoneme sequences.

Next, a lexicon optimization method based on the CM is proposed. Section 12.4 evaluates the performance of an automatic speech recognition (ASR) system employing the proposed method in terms of computational complexity and WERs. Finally, the findings are summarized in Sect. 12.5.

12.2 G2P Model-Based Lexicon Generation

G2P conversion is used to predict phoneme sequences by aligning the graphemes of words or sentences with phonemes (Bisani and Ney 2008). One of the simplest G2P conversions is achieved by a dictionary lookup (Bisani and Ney 2008). That is, for a given input grapheme sequence, a possible phoneme sequence is obtained with a look-up table. Therefore, the dictionary look-up approach is time-consuming and tedious. Moreover, it is difficult to find the pronunciation of unseen words in this way, because the dictionary used for the lookup is finite. In addition, it does not enable unseen words to be found that do not exist in the dictionary.

To overcome the limitations of such finite dictionaries, a data-driven approach is used for the G2P conversion (Bisani and Ney 2008). This is usually performed by mapping 1 to N-best after designing a joint-sequence model from a training corpus. Figure 12.1 shows an example of the G2P conversion for the given word "JOINT." As shown in the figure, this word can be represented by three different phoneme sequences.

12.3 Proposed Lexicon Optimization

CM can be defined by the linguistic distance between two phoneme sequences in the expanded lexicon of a G2P model (Bisani and Ney 2008). In this section, we propose a lexicon optimization method that is defined by the acoustic distance between two phoneme sequences using inter-phone and inter-word distances. The proposed method is explained in detail in the following subsections.

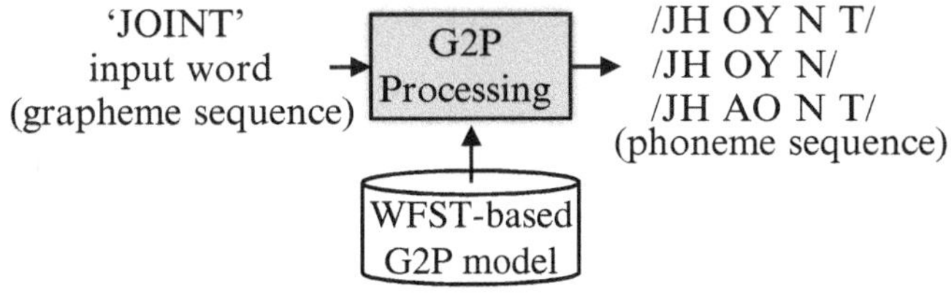

Fig. 12.1 Example of G2P conversion for given word "JOINT"

12.3.1 Confusability Measure Using Distance between Phoneme Sequences

First, let W_i be the ith word in the original N-best lexicons from the G2P conversion that has phoneme sequences, and let $s_{i,j}$ $(j = 1, \ldots, N)$ be the 1-to-N-best mapped phoneme sequences. Then, the CM of $s_{i,j}$ is defined as (Kim et al. 2008)

$$\mathrm{CM}\left(s_{i,j}\right) = L\left(s_{i,j}\right) \min_{\substack{1 \le k \le N_W,\, k \ne i \\ 1 \le l \le N}} \left(D\left(s_{i,j}, s_{k,l}\right) L\left(s_{k,l}\right)\right), \tag{12.1}$$

where N_W is the number of words and $D(x, y)$ is the dynamic programming (DP)-based phoneme sequence distance that is defined by the HMM-based phoneme distance (Anguita et al. 2005). In addition, $L(x)$ is defined as the normalized length by $l_{\max}$ as

$$L(x) = \frac{\#(x)}{l_{\max}}, \tag{12.2}$$

where $\#(x)$ is the number of phonemes in x and $l_{\max} = \max_{1 \le i,j \le N} \#\left(s_{i,j}\right)$ is the maximum length in the N-best G2P-converted lexicon.

12.3.2 Phoneme Sequence Distance Measure

12.3.2.1 HMM-Based Phoneme Distance Measure

The acoustic distance between two phonemes can be calculated by using acoustic models (Anguita et al. 2005), and it is defined as

$$d_{\mathrm{HMM}}\left(p_1, p_2\right) = \frac{\sum_Q P(Q) \frac{1}{L} \sum_{i=1}^{L} D_N\left(N_{q_{1i}}, N_{q_{2i}}\right)}{\sum_Q P(Q)}, \tag{12.3}$$

where Q is the alignment between the HMM states of the phones p_1 and p_2, $P(Q)$ is the probability of Q, L is the length of the alignment, q_{1i} and q_{2i} are the states of the models that are aligned according to Q, $N_{q_{1i}}$ and $N_{q_{2i}}$ are the Gaussian distributions associated with the states q_{1i} and q_{2i}, and $D_N(\,)$ is the distance between the two Gaussian distributions. In Eq. (12.3), $P(Q)$ is calculated by multiplying the transition probabilities of both phoneme state sequences. Figure 12.2 shows an example of possible $P(Q)$'s that are represented as $(q_{11} \rightarrow q_{12}, q_{21} \rightarrow q_{22})$, $(q_{12} \rightarrow q_{12}, q_{22} \rightarrow q_{23})$, and $(q_{12} \rightarrow q_{13}, q_{33} \rightarrow q_{23})$.

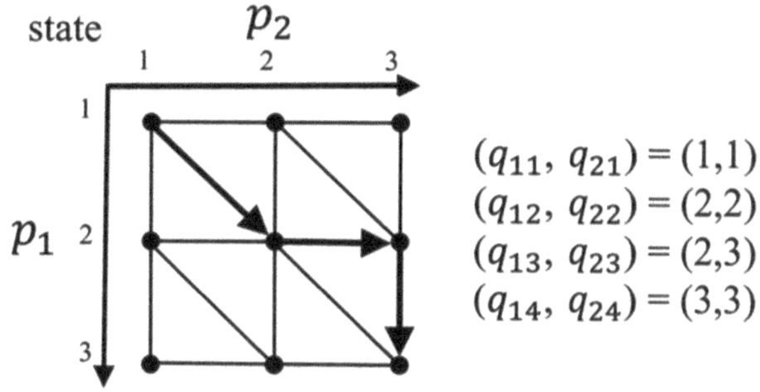

Fig. 12.2 Subset of alignments used to calculate inter-HMM distance, where bold lines correspond to alignment Q in Eq. (12.3). The values of q_{1i} and q_{2i} are aligned states (Anguita et al. 2005)

The acoustic model for calculating the distance between the two phonemes can be represented as one Gaussian distribution for each state of the HMM models (Sooful and Botha 2001). In this paper, we calculated $D_N(\cdot)$, using each of the three different distance measures such as Euclidean (EUC) distance, Mahalanobis (MAH) distance, and symmetric Kullback–Leibler (KL) distance (Sooful and Botha 2001).

12.3.2.2 DP-Based Phoneme Sequence Distance Measure

A dynamic time warping (DTW) technique is incorporated into the acoustic distance to determine how different the two phoneme sequences are. The DTW is defined as (Anguita et al. 2005)

$$D(x, y) = d_{\mathrm{DTW}}\left(s_x, s_y\right), \tag{12.4}$$

where

$$d_{\mathrm{DTW}}\left(s_x, s_y\right) = \min_F \left[\frac{\sum_{k=1}^{K} d_{\mathrm{HMM}}\left(p_x(k), p_y(k)\right) w(k)}{\sum_{k=1}^{K} w(k)} \right]. \tag{12.5}$$

In Eq. (12.5), $d_{\mathrm{HMM}}(p_{1i}, p_{2j})$ is the distance between the HMMs described in Eq. (12.3), and the weighting function, $w(k)$, applied to the DTW distance is used to normalize the path F and is defined as (Anguita et al. 2005)

$$w(k) = i(k) - i(k-1) + j(k) - j(k-1), \tag{12.6}$$

where $i(1) = j(1) = 0$. In addition, $c(k)$ in the path $F = \{c(1), c(2), \ldots, c(K)\}$ consists of the pair of coordinates $(i(k), j(k))$ in the i and j directions, respectively, when K is the number of alignments of the two phoneme sequences.

The measure obtained with DTW is the minimum weighted sum of the distance between the phoneme sequences for all the possible alignments between the sequences. Therefore, the DTW technique forces an alignment that minimizes the accumulated distance and forces the two sequences to consider the similarity.

Table 12.1 Example of CM scores for phoneme sequences of the word "STATUE" obtained by 1-to-4-best mapping

4-Best phoneme sequence	CM score
S T AE CH UW	0.0497
S T AE CH Y UW	0.0499
S T AE CH UW EH	0.0190
S T AE CH UW AH	0.0181

12.3.3 Lexicon Optimization Using CM

In this subsection, we describe how to optimize the lexicon using a CM. The proposed method selects the phoneme sequences with CM scores higher than a predefined threshold, except one phoneme sequence for each word in the original lexicon that has the highest CM score first maintained in the optimized lexicon. Next, the phoneme sequences with CM scores lower than the threshold are assumed to be confusable words and will not appear in the pruned lexicon.

Table 12.1 provides an example of the phoneme sequences obtained by the 1-to-4-best G2P conversion for the word "STATUE" and their CM scores. In this case, the most probable phoneme sequence is /S T AE CH UW/. If the threshold is 0.02, two phoneme sequences, /S T AE CH UW/ and /S T AE CH Y UW/, will remain in the lexicon.

12.3.4 Decoding Network Generation

A WFST-based decoder for LVCSR is fully composed as $H \circ C \circ L \circ G$, where four different WFSTs—H, C, L, and G—represent the HMM state level topology, the context dependency expansion, the lexicon, and the n-gram LM, respectively (Mohri et al. 2008). Therefore, the proposed lexicon optimization method transforms the lexicon, L, into the optimized lexicon, L'. Thus, we obtain the WFST-based decoder composed as $H \circ C \circ L' \circ G$.

12.4 Speech Recognition Experiment

To evaluate the performance of the lexicon optimization method, we constructed the following ASR systems: a baseline ASR system (Baseline), an ASR system of a 1-to-4-best G2P-converted lexicon, and ASR systems based on lexicons pruned by the proposed lexicon optimization method using different acoustic distances. The baseline system was constructed by the Kaldi speech recognition toolkit (Povey et al. 2011) with 7138 utterances from *The Wall Street Journal* (WSJ0) (Paul and Baker 1992). In addition, for the baseline lexicon, a 1-best G2P lexicon was used. As a feature of the system, 39-dimensional mel-frequency cepstral coefficients

Table 12.2 Performance evaluation of an LVCSR system employing the proposed method

ASR system	Baseline	4-Best G2P converted lexicon	Proposed method		
			EUC	KL	MAH
WER (%)	12.19	13.93	12.58	12.42	11.88
RTF	0.239	0.476	0.364	0.381	0.380

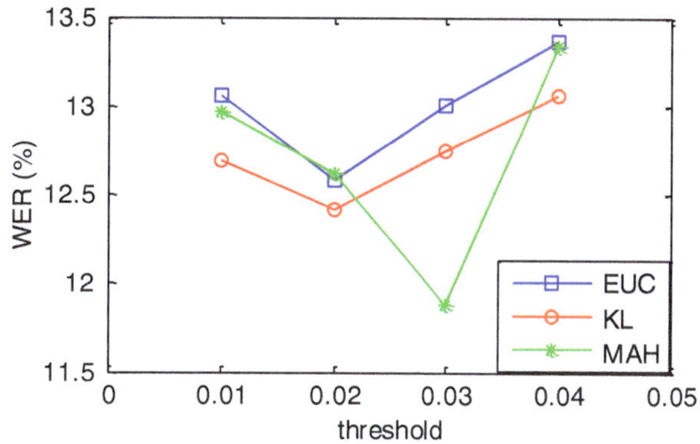

Fig. 12.3 Comparison of word error rates of an ASR system employing different distance measures depending on the threshold

(MFCCs) were used, and the cepstral mean normalization (CMN) was applied to the feature vector. The acoustic model was constructed by means of concatenating context-dependent HMMs, and a trigram LM was constructed from a set of sentences from the WSJ0 with a vocabulary of 20k different words. The test sub-corpus was also extracted from the WSJ0 and was composed of 333 utterances containing 5643 different words.

Table 12.2 compares the WER and real-time factor (RTF) for each ASR system using a lexicon obtained from the 1-best G2P-converted lexicon, a 1-to-4-best G2P-converted lexicon, a pruned lexicon based on the proposed method with different phoneme distances, and EUC, KL, and MAH distances (Sooful and Botha 2001). As shown in the table, the RTF and WER were lowered with the different phoneme distances.

Next, we evaluated the performance of the proposed method by changing the threshold from 0.01 to 0.04 at a step of 0.01. As shown in Fig. 12.3, the average WER was lowered. However, as the threshold became greater than 0.02, the average WER of the proposed method also increased. This was because the phoneme sequences were pruned excessively. Consequently, by applying the proposed method with MAH, we could achieve a relative WER reduction of 14.72 % compared to that achieved with a lexicon of a 1-to-4-best G2P conversion.

12.5 Conclusion

In this paper, we proposed a lexicon optimization method based on a CM to reduce the decoding network of lexicons constructed by the G2P model. When the lexicon was built to find the phoneme sequences of unseen words, the lexicon often became oversized, causing an increase in the size of the LVCSR decoder. Consequently, the performance of the LVCSR was lowered. On the other hand, the proposed lexicon optimization method was used to reduce the decoding network by pruning phoneme sequences that were much more confusable than others. It was shown from the ASR experiments that an ASR system employing a lexicon optimized by the proposed method provided a relative WER reduction of 14.72 % compared to that of a lexicon from a 1-to-4-best G2P conversion.

Acknowledgments This work was supported in part by the National Research Foundation of Korea (NRF) grant funded by the government of Korea (MSIP) (No. 2015R1A2A1A05001687) and by the MSIP, Korea, under the ITRC (Information Technology Research Center) support program (IITP-2015-H8501-15-1016) supervised by the IITP (Institute for Information & communications Technology Promotion).

References

Anguita J, Hernando J, Peillon S, Bramoulle A (2005) Detection of confusable words in automatic speech recognition. IEEE Signal Process Lett 12(8):585–588

Bisani M, Ney H (2008) Joint-sequence models for grapheme-to-phoneme conversion. Speech Commun 50(5):434–451

Guo Y, Li T, Si Y, Pan J, Yan Y (2012) Optimized large vocabulary WFST speech recognition system. In: Proceedings of FSKD, Chongqing, China, pp 1243–1247

Jitsuhiro T, Takahashi S, Aikawa K (1998) Rejection of out-of-vocabulary words using phoneme confidence likelihood. In: Proceedings of ICASSP, Seattle, WA, pp 217–220

Kanthak S, Ney H, Riley M, Mohri M (2002) A comparison of two LVR search optimization techniques. In: Proceedings of Interspeech, Denver, CO, pp 1309–1312

Kim MA, Oh YR, Kim HK (2008) Optimizing multiple pronunciation dictionary based on a confusability measure for non-native speech recognition. In: Proceedings of IASTED, Innsbruck, Austria, pp 215–220

Kubo Y, Watanabe S, Nakamura A (2012) Decoding network optimization using minimum transition error training. In: Proceedings of ICASSP, Kyoto, Japan, pp 4197–4200

Lin S, Yvon F (2007) Optimization on decoding graphs by discriminative training. In: Proceedings of Interspeech, Antwerp, Belgium, pp 1737–1740

Mohri M, Pereira F, Riley M (2008) Speech recognition with weighted finite-state transducers. In: Handbook on speech processing and speech communication. Springer, Berlin, pp 559–582

Neukirchen C, Willett D, Rigoll G (1997) Reduced lexicon trees for decoding in a MMI Connectionist/HMM speech recognition system. In: Proceedings of Eurospeech, Rhodes, Greece, pp 2639–2642

Paul DB, Baker JM (1992) The design for the Wall Street Journal-based CSR corpus. In: Proceedings of ICSLP, Stroudsburg, PA, pp 357–362

Povey D et al (2011) The Kaldi speech recognition toolkit. In: Proceedings of ASRU, Honolulu, HI, pp 1–4

Saon G, Chien J-T (2012) Large-vocabulary continuous speech recognition systems – a look at some recent advances. IEEE Signal Process Mag 29(6):18–33

Sooful JJ, Botha EC (2001) An acoustic distance measure for automatic cross-language phoneme mapping. In: Proceedings of PRASA, Franschhoek, South Africa, pp 99–102

Vinyals O, Deng L, Yu D, Acero A (2009) Discriminative pronunciation learning using phonetic decoder and minimum-classification-error criterion. In: Proceedings of ICASSP, Taipei, Taiwan, pp 4445–4448

Chapter 13
Linguistic Individuality Transformation for Spoken Language

Masahiro Mizukami, Graham Neubig, Sakriani Sakti, Tomoki Toda, and Satoshi Nakamura

Abstract In text and speech, there are various features that express the individuality of the writer or speaker. In this paper, we take a step towards the creation of dialogue systems that consider this individuality by proposing a method for transforming individuality using a technique inspired by statistical machine translation (SMT). However, finding a parallel corpus with identical semantic content but different individuality is difficult, precluding the use of standard SMT techniques. Thus, in this paper, we focus on methods for creating a translation model (TM) using techniques from the paraphrasing literature and a language model (LM) by combining small amounts of individuality-rich data with larger amounts of background text. We perform an automatic and manual evaluation comparing the effectiveness of three types of TM construction techniques and find that the proposed system using a method focusing on a limited set of function words is most effective and can transform individuality to a degree that is both noticeable and identifiable.

Keywords Linguistic individuality • Statistical machine translation • Paraphrasing

13.1 Introduction

In language, the words chosen by the speaker or writer transmit not only semantic content but also other information such as aspects of their individuality, personality, or speaking style. While not directly related to the message, these aspects of language are extremely important to achieve rapport between the person creating the message and its intended target. We can assume that this observation will also carry over to human computer interaction (Metze et al. 2009).

For example, in a situation where a dialogue system is used to represent famous characters in movies or comics, we would like to reproduce the character's

M. Mizukami (✉) • G. Neubig • S. Sakti • T. Toda • S. Nakamura
NAIST, 8916-5 Takayama-cho, Ikoma, Nara 630-0192, Japan
e-mail: masahiro-mi@is.naist.jp; neubig@is.naist.jp; ssakti@is.naist.jp; tomoki@is.naist.jp; s-nakamura@is.naist.jp

G. Geunbae Lee et al. (eds.), *Natural Language Dialog Systems and Intelligent Assistants*, DOI 10.1007/978-3-319-19291-8_13

well-known and unique expressions. It is also natural that a dialogue system can realize smoother communication by talking in a more polite way to adults and a more friendly and informal way to children (Wang et al. 2012). To make these sorts of applications possible, the ability to express a rich variety of individuality and atmosphere depending on the type of user or scene is neccessary (Isard et al. 2006; Mairesse and Walker 2011).

In this paper, we define the individuality as the elements which allows us to distinguish unique person from other person. Individuality is closely related to personality, and previous work has modeled personality using measures such as the Big Five Traits (Gosling et al. 2003). Previous work has also noted that coherence of acoustic and linguistic traits has a strong influence on perceptions of individuality (Isbister and Nass 2000).

Handling of individuality of features of the voice (i.e., "acoustic individuality") is a widely researched topic in speech synthesis and translation (Abe et al. 1988; Yamagishi et al. 2010; Qian et al. 2013). On the other hand, there are few studies that attempt to control the individuality of each speaker as expressed on the lexical level through choice of words or expressions, etc. (i.e., "linguistic individuality"). There are some works that attempt to generate sentences that express a certain personality based on rule-based sentence generation (Mairesse and Walker 2011), personality infused n-gram models (Isard et al. 2006). However, while controlling personality is certainly a first step in the direction of creating a richer user experience, research in the area overall is sparse, and controlling personality will not allow us to, for example, reproduce the unique expressions of a single speaker.

In this paper, we propose a technique that takes text as input and converts the text into text that reflects the individuality of a target speaker. This approach has two differences from the previously mentioned work on personality-sensitive natural language generation. The first is that our method handles not *generation*, but *transformation*, taking as input a natural language sentence and converting the individuality of the source speaker into that of the target speaker. This has the advantage that it can be used as a post-processing step either for dialogue systems where generation is used as a black box or for other applications that do not explicitly use generation, such as machine translation. In addition, by focusing not on *personality*, but *individuality*, we are able to cover applications such as the previously mentioned dialogue system mimicking a famous character.

We propose the probabilistic framework for transforming individuality like statistical machine translation. This framework is based on previous work (Xu et al. 2012; Brill and Moore 2000; Neubig et al. 2012) that uses machine translation techniques to translate betweens speaking or writing styles. However, in contrast to these works, which rely on parallel data of the source and target styles, it is difficult to prepare a large quantity of parallel data between source and target speakers for individuality translation.

In this framework, we define a translation model (TM) that has the ability to translate between individualities of speakers and a language model (LM) that reflects the individuality of the target speaker. For the LM, we use a small collection of text created by the target speaker and a larger background model. For the TM, as it is difficult to create the parallel data necessary to train standard MT systems, we examine techniques from the paraphrasing literature, acquiring paraphrases using a thesaurus, distributional similarity, and bilingual parallel text.

Based on the results of the analysis, we find that in the system proposed in this paper, conversion of function words allows for detectable and identifiable increases in the individuality of the target sentence. On the other hand, conversion of content words is less successful, leaving important challenges for future work.

13.2 A Probabilistic Framework for Transforming Individuality

In this section, we describe our proposed method for translation of speaker individuality. To create a method capable of this conversion, we build upon previous work that has studied conversion of writing or speaking style (Xu et al. 2012; Brill and Moore 2000; Neubig et al. 2012).

Specifically, we build upon the work of Neubig et al. (2012), which was originally conceived for translation from spoken to written text or for translation of text from one style to another. Given a string of input words V (representing a spoken language sentence) and a string of words W (representing a written language sentence), we transform V to W using the noisy channel model. In consideration of the quantity of available corpora, the posterior probability $P(W|V)$ is decomposed into TM probability $P(V|W)$, which must be estimated from a corpus of parallel sentences, which is more difficult to find, and LM probability $P(W)$, which can be estimated from a corpus of only output side text which we can secure in large quantities:

$$P(W|V) = \frac{P(V|W)P(W)}{P(V)}. \tag{13.1}$$

Given this probabilistic model, the output is found by searching for the output sentence $\hat{W}$ that maximizes $P(W|V)$. $P(V)$ is not affected by choice of W, so this maximization is expressed as follows:

$$\hat{W} = \underset{W}{\operatorname{argmax}}\, P(V|W)P(W). \tag{13.2}$$

In addition, because the LM probability $P(W)$ tends to prefer shorter sentences, we also follow standard practice in machine translation (Och and Ney 2002) in introducing a word penalty proportional to sentence length $|W|$. We combine these three elements in a log-linear model, with parameters λ_{tm}, λ_{lm}, and λ_{wp} as follows:

$$\hat{W} = \underset{W}{\operatorname{argmax}}\, \lambda_{tm} \log P(V|W) + \lambda_{lm} \log P(W) + \lambda_{wp}|W|. \tag{13.3}$$

Following this framework, we consider a setting in which we translate from utterance V that expresses the individuality of the source speaker to utterance W that expresses the individuality of target speaker. However, compared to the previously mentioned style transformation or standard SMT, we are faced with a drastic lack of data. The amount of target side data W is limited, and we will often have no parallel data with identical semantic content expressed with the individuality of the target and source speakers. In fact, when we had one author of the paper attempt to make this data in preliminary experiments, we found that even when an annotator is available, creation of the data is quite difficult and time-consuming. If the annotator attempted to follow the semantic content of the input faithfully, it was difficult to express a rich variety of individuality, and when the annotator attempted to edit more freely, the individuality was expressed abundantly, but in many cases the semantic content changed too much to be used reliably as training or testing data for the system.

In the next two sections, we describe how we build a system even in situations where no parallel data is available to train that TM probability $P(V|W)$.

13.3 Language Model

For transforming individuality, it is necessary to build an LM that express the individuality of the target speaker.

13.3.1 Language Model Training

For transforming individuality, it is necessary to build an LM that expresses the individuality of the target speaker. In order to do so, we need to collect data that expresses the target speaker's speaking style. In addition, it is better if the data used to train the LM matches the content of the data to be converted. Thus, an initial attempt to create an LM that expresses the speaking style of the target will start with gathering data from the speaker and training an n-gram LM on this data.

13.3.2 Language Model Adaptation

When we collect the utterance of only one target speaker and build an LM, it is difficult to collect a large number of utterances from any one speaker. Thus the contents covered by the LM are restricted. Therefore, an LM made with only data from the target speaker cannot estimate the LM probability $P(W)$ accurately. To remedy this problem, in this paper we build a target LM that interpolates a small LM $P_t(W)$ that is trained as explained in the previous section and an LM $P_g(W)$ that is trained from a large-scale corpus. Using an interpolation coefficient λ, we combine these two models using linear interpolation

$$P(W) = \lambda P_t(W) + (1 - \lambda)P_g(W). \tag{13.4}$$

We calculate λ to generate LM $P(W)$, such that we achieve the maximum LM probability on a held out development set also created using data from the target speaker. Note that this framework is flexible, so we could also add an additional LM considering the personality of the speaker (Isard et al. 2006), but in this paper for simplicity we only use two models: the general domain and with the target speaker's individuality.

13.4 Translation Model

Now that we have modeled individuality in the LM, we must next create a translation model $P(V|W)$ that expresses the possible transformations changing the style, but not the semantic content, of the utterance. However, as mentioned in Sect. 13.2, it is nontrivial to collect a corpus of sentences spoken by the source and target speaker while having the same meaning, so we will have to create this model without relying on a parallel corpus.

In this paper, we solve this problem by building the TM using techniques from paraphrasing. In this work, we focus on methods for paraphrasing using a thesaurus, n-gram-based distributional similarity, and bilingual parallel text, with each of the three resources playing a different role.

13.4.1 Translation Model Using Thesauri

Thesauri are language resources specifying groups of synonyms and thus are a good resource for reliably finding semantically plausible transformations. The most widely used thesaurus in the NLP community is Wordnet (Miller 1995), and its counterpart in Japanese, our target language, is Japanese Wordnet (Bond et al. 2009). The TM built using a THESAURUS is used to find replacement candidates based on synonyms for nouns and verbs, similarly to previous works on paraphrasing using thesauri (Inui and Fujita 2004).

Using this thesaurus, we build the TM according to the following procedure.

1. For each word in the input, search the WordNet with the word as the query.
2. When the word is found, acquire all synonyms from WordNet using the synset.
3. Calculate the TM probability (Sect. 13.4.3) for all words and store them in the TM.

We show an example of the TM acquired by this method in Table 13.1.

Table 13.1 A sample of the TM using thesauri

Source	Target	TM prob.
カメラ (camera)	カメラ (camera)	0.95
	キャメラ (kamera)	0.01
	ビデオカメラ (video camera)	0.01
	写真機 (photo machine) and other 2 words	0.01
良い (good)	良い (good)	0.4
	いい (nice)	0.4
	よろしい (fine)	0.01
	見事 (excellent) and other 42 words	0.01

13.4.2 Translation Model Using Distributional Similarity

Thesauri have the advantage of providing broad coverage, but they also consist mainly of synonyms for nouns and verbs and don't have data regarding synonymy of fillers, exclamations, particles, and other function words. However, these elements are very important in expressing a number of aspects of language (Chung and Pennebaker 2007). Especially in Japanese, particles at the end of the sentence and auxiliary verb particle have been noted as playing an important role in expressing individuality (Teshigawara and Kinsui 2012).

The TM is built according to the following procedure.

1. Prepare a list of function words by performing POS tagging on the training corpus and extracting all non-content words.
2. Count all 3-g in the target speaker's utterances.
3. Find groups of *n*-grams that have a function word in the second position and the same first and third words, and add groups to the set of potential synonyms (e.g., that's *so* great, that's *really* great).
4. Calculate the TM probability for all words and store them in the TM.

We show an example of a TM acquired by this method in Table 13.2. We can extract non-content word and particle paraphrases. In this method, we don't consider meaning of words, and we sometime get wrong paraphrases of the meaning, for example, "it *for* you" and "it *from* you." We check this problem by evaluating transforming word error rate (WER).

Table 13.2 A sample of the TM using n-grams

Source	Target	TM prob.
です (is)	です (is)	0.7
	だ (is: informal)	0.3
けど (but)	けど (but)	0.8
	よ (yes)	0.2
も (also)	も (also)	0.6
	で (at)	0.4
が (SUBJ)	が (SUBJ)	0.6
	は (SUBJ)	0.4

13.4.3 Calculation of Translation Model Probability

While the two previous methods can find potential candidates for translation, it gives us no mechanism to determine how reliable these candidates are. However, we also found in preliminary experiments that simply assigning a uniform probability to all transformations in the previous sections was not sufficient to accurately decide when words are interchangeable. To solve this problem, we calculate TM probabilities using n-gram similarity.

We base our method on techniques to acquire synonyms from nonparallel corpora (Dagan et al. 1999; Barzilay and Lee 2003). In the previous works, similarity of the word itself is calculated from a nonparallel corpus according to the contextual similarity of the words.

In order to calculate this contextual similarity, we prepare a bigram LM with vocabulary L and decide the similarity $\mathrm{Sim}(w, v)$ for two words w and v as follows:

$$\mathrm{Sim}(w, v) = 1 - \frac{1}{2|L|}\left(\sum_{l \in L}\left|P(w|l) - P(v|l)\right| + \sum_{l \in L}\left|P(l|w) - P(l|v)\right|\right). \tag{13.5}$$

Similarity $\mathrm{Sim}(w, v)$ is decided by the similarity of n-gram distributions, based on the distributional hypothesis that words that appear in similar contexts have a similar role. For the calculated similarity $\mathrm{Sim}(w, v)$, we normalize over values of $\mathrm{Sim}(w, v)$ for all words, so that the probabilities sum to one

$$P(w|v) = \frac{\mathrm{Sim}(w, v)}{\sum_{L \in l} \mathrm{Sim}(l, v)}. \tag{13.6}$$

Thus, we can approximate TM probability of words w and v without using a parallel corpus.

13.4.4 Translation Model Using Bilingual Text

The final method we examine for creating the TM is based on Bannard and Callison-Burch (2005)'s method for using bilingual text to train a paraphrasing model. Paraphrases acquired by this method have the advantage of providing broad coverage (theoretically it is possible to cover both content and function words) and allowing for acquiring of multi-word transformations.

Assume we have two phrases v and w in the language under consideration (in our case, Japanese) and also have a phrase-based TM indicating the translation probabilities to and from a phrase e in a different language (in our case, English). We decide the paraphrase probability $P(w|v)$ using translation probabilities $P(w|e)$ and $P(e|v)$ by using the English phrase e as a pivot as follows:

$$P(w|v) = \sum_e P(w|e)P(e|v). \tag{13.7}$$

The TM probabilities can be computed using standard methods from SMT (Koehn et al. 2003). The details of the phrase table that we used in the construction of paraphrases for this work are shown in Table 13.3.[1]

We show an example of a TM acquired by this method in Table 13.4.

Table 13.3 The details of the phrase table

Corpus	BILINGUAL corpus including Wikipedia, lecture, newspaper, magazine and dialogue
Words	24.2M (en)
	29.6M (ja)
Phrases	67.1M
Max length	7 words
Alignment	Nile (Riesa et al. 2011)
Parsing	Kytea (Neubig et al. 2011)

Table 13.4 A sample of paraphrase acquired from bilingual data for "翻訳された(translated)"

Translation	TM prob.
翻訳された (translated)	0.083
に翻訳された (translated to)	0.034
翻訳 (translate)	0.012
共訳 (joint translation)	0.011
訳される (was translated)	0.011
と訳された (was translated to)	0.002
and 20 other phrases	

[1]This Japanese paraphrase model will be made available upon acceptance of the paper.

13.5 Evaluation Measures for Individuality Transformation

In previous work, they evaluate the relationship between some automatic evaluation metrics and various human judgments. Automatic metrics based on LMs have better correlation with human judgments than existing metrics in the context of previous work. We evaluate our proposed method under the same conditions as previous work's automatic evaluation metrics as LM.

In manual evaluation, they evaluate based on human judgments of semantic adequacy, lexical dissimilarity, and stylistic similarity, because they clarify style and the relations with individual elements. We consider it, propose several evaluation measures for transforming of individuality that focus on *individuality* of the target speaker, *accuracy* of conversion, and *breadth* of possible conversion.

13.5.1 Automatic Evaluation

In automatic evaluation, we use the two following measures.

LM Ratio Xu et al. (2012) proposed a method for evaluating the style of a converted sentence using the ratio of language model probabilities, where P_t is the probability of a model trained on target domain data and P_s is the probability of a source domain language model:

$$P(\text{style} = \text{target}|\text{sentence}) = \frac{P_t(\text{sentence})}{P_s(\text{sentence}) + P_t(\text{sentence})}. \quad (13.8)$$

Coverage We define coverage as the ratio of words for which there is a conversion candidate in the TM. A TM that can convert various vocabulary will have a higher coverage, and thus coverage can be used to evaluate the breadth of the conversion.

13.5.2 Manual Evaluation

While automatic evaluation is useful for the rapid development of systems, it is difficult to evaluate small differences in nuance. Thus, we also perform manual evaluation to evaluate correctness and individuality of the output. Specifically, we evaluate two following factors.

Individuality In order to evaluate individuality, we first have a subject read the training data to learn the individuality of target speaker. The subject is then shown the system output and asked "does this sentence reflect the individuality of person who wrote the training data?" The subject then assigns a score of 1 (do not agree) to 5 (do agree).

Word Error Rate: This is the ratio of words converted by our method that are syntactically or semantically incorrect in the post-conversion sentence. This is calculated by having the subject look at the sentence before and after conversion and point out conversion mistakes.

13.6 Experimental Evaluation

In order to evaluate the proposed method, we performed an evaluation focused on how well the proposed model can reproduce the individuality of a particular speaker.

13.6.1 Experiment Conditions

As data for our research, we use a camera sales dialogue corpus (Hiraoka et al. 2014) that consists of one-on-one sales dialogues between three salesclerks and 19 customers. We split the corpus of three salesclerks into one corpus for every speaker each and further divide each of these corpora into training, development, and evaluation data. The details of the data for each of the salespeople are shown in Table 13.5. All conversations were performed in Japanese by native or highly fluent Japanese speakers. As mentioned in Sect. 13.3.2, in order to create an LM that is both sufficiently accurate and expresses the personality of the speaker, we use multiple LMs created using data from the target speaker and a larger background corpus. As our target speaker data, we use the training data from the previously described camera sales corpus. As our large background corpus, we use data from the BTEC (Takezawa et al. 2002) and the REIJIRO[2] dictionary example sentence corpus. The size of these background corpora is also shown in

Table 13.5 Number of utterances and words in the camera sales dialogue corpus

	Clerk	Utterance	Word
Train	A	238	11,758
	B	240	12,495
	C	228	9039
Develop	A	65	3016
	B	43	2271
	C	37	1462
Test	A	9	173
	B	9	134
	C	9	148

[2]http://www.eijiro.jp.

Table 13.6 Number of sentences and words in BTEC and REIJIRO

Corpus	Sentence	Word
BTEC	465k	4.11M
REIJIRO	424k	8.90M
SUM	889k	13.01M

Table 13.6. We calculate the linear interpolation parameter to maximize likelihood on the development data. As a result, the linear interpolation parameter λ became 0.88. For the log-linear model in Eq. (13.3), we set $\lambda_{tm} = \lambda_{lm} = 1$ and adjust the word penalty so that the length of sentences before and after transformation is approximately equal.[3]

We perform an evaluation over three combinations of TMs for conversion of individuality. We compare the three methods for constructing the TM using the thesaurus (THESAURUS), *n*-gram similarity (SIMILARITY), and parallel corpus (BILINGUAL). We also compare with a baseline method that does not perform any conversion at all (SOURCE).

In the experimental evaluation, we first have subjects read the training data of the target speaker. Next, we prepare an input sentence that is selected randomly from other salesclerks. Based on this input sentence, we use the three methods described in the previous paragraph to convert it into the target speaker's individuality. The subject reads these three results. The subject estimates WER and individuality for each of these four conversion results according to the measures described in Sect. 13.5.2, and we also automatically calculate LM measure and coverage according to the measures described in Sect. 13.5.1.

In this evaluation, three subjects evaluate result for three speakers, each with nine utterances, 27 conversion results in total. We find the confidence interval of each evaluation measure using bootstrap resampling (Koehn 2004) with significance level $p < 0.05$.

13.6.2 Experiment Result

In this section, we describe the results of our evaluation of the proposed method for transforming individuality of text. We first discuss the results for the automatic evaluation measures. In Fig. 13.1 we show the coverage, in Fig. 13.2 we show the LM ratio.

In this evaluation, when we used BILINGUAL, coverage improved most, with a total of 80 % of words being possible candidates for replacement. However, when we used BILINGUAL, the percentage of words changed was 7.6 %, lower than that of SIMILARITY, with a total of 13.0 %. This is because function words (acquired by SIMILARITY) are more easily replaceable than content words or mixed

[3]Verbosity is one component of individuality, so setting λ_{wp} to a different value for each source/target speaker pair is more appropriate, but we leave this to future work.

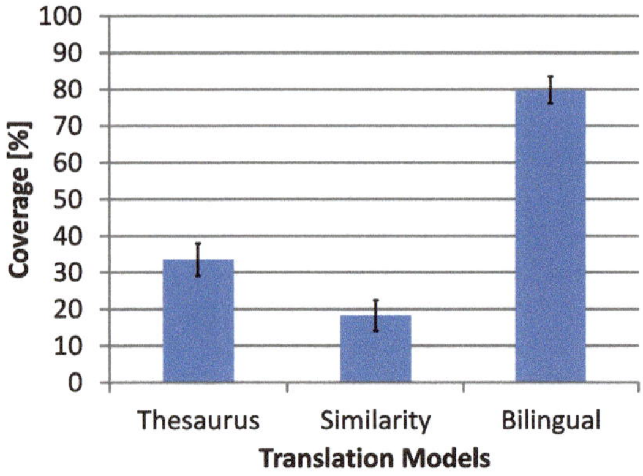

Fig. 13.1 Coverage for each model

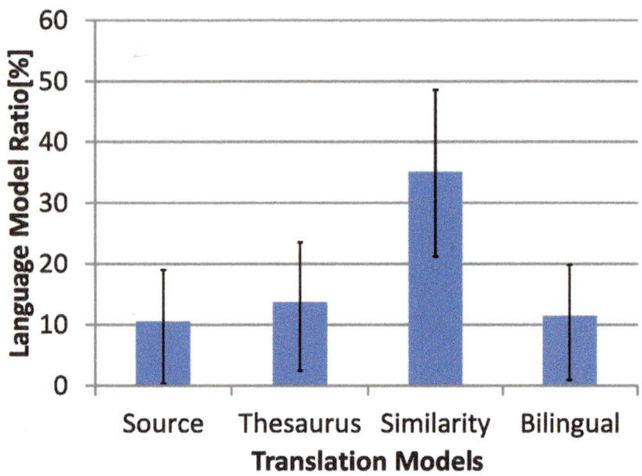

Fig. 13.2 Language model ratio for each model

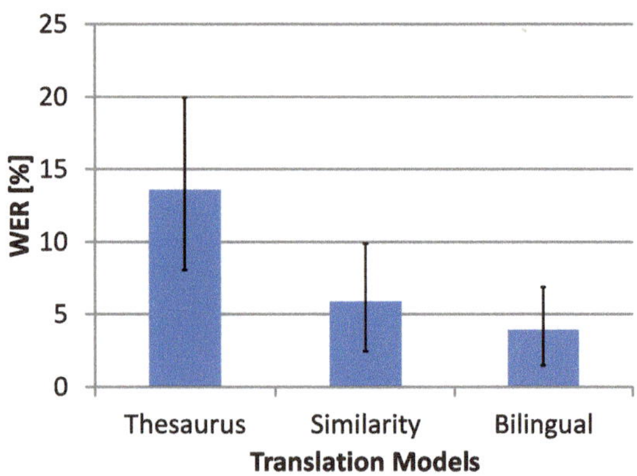

Fig. 13.3 WER for each model

phrases containing both function and content words. In addition, when we used the SIMILARITY TM, LM ratio improved most, with a total of 35 % from 10 % in SOURCE.

We show the results of manual evaluation of WER in Fig. 13.3 and individuality in Fig. 13.4. The first result to be noted is that transformation using SIMILARITY is

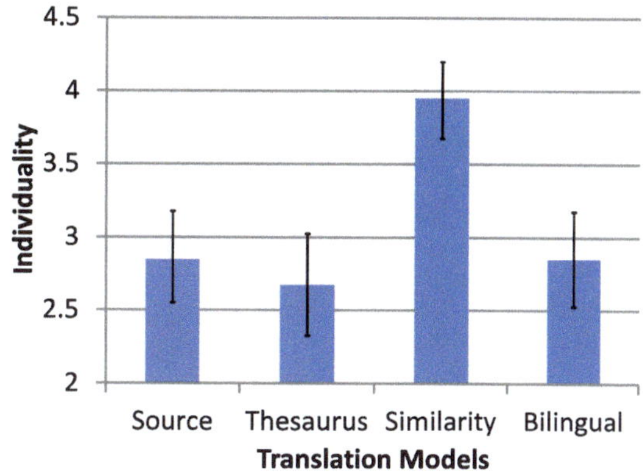

Fig. 13.4 Individuality for each model

able to raise the individuality to 3.9 from the SOURCE of 2.8, a significant difference. If we compare it with Fig. 13.4, LM ratio and individuality understand a similar thing evaluating. This demonstrates that our proposed method of transforming the individuality of speakers is able to successfully do so to a noticeable degree.

However, the results given the other two methods were mixed. For THESAURUS, we can see the individuality actually unchanged. This is due to the fact that the WER of this method was high, and often the meaning of the sentence was lost due to mistaken conversions of content words. This unnaturalness resulted in very low evaluations of individuality.

When we used BILINGUAL, coverage and WER generally improved, and change rate improved over THESAURUS (which was 4.3 %). However, LM ratio and individuality didn't improve over the SOURCE. The reason for this is that function words are very important for expressing individuality (Chung and Pennebaker 2007; Teshigawara and Kinsui 2012), but were not well enough represented in the paraphrases acquired from bilingual data. The reason for this is twofold. First, data that is translated between languages usually contains few fillers (as they are deleted before translation) and other common spoken expressions. Second, in our case we used English and Japanese for the pivot languages, but these languages diverge in their use of function words (for example, English does not use explicit case markers, and Japanese does not use articles), making acquiring good transformations for these words difficult. This is illustrated by the fact that BILINGUAL only covers a total of 11 % of the transformations covered by SIMILARITY.

13.7 Conclusion

In this paper, we proposed a method for transforming individuality. We performed an evaluation of the effectiveness of TMs acquired using n-gram similarity, thesauri, and bilingual text in this context. We found that function word transformations based on n-gram similarity were the most effective in improving the individuality of text.

While the experimental results showed that the proposed technique is able to successfully convert speaker individuality to some extent, there are still a number of future challenges related to refining the language and TMs to convert speaker individuality more precisely. The main area for improvement lies in improvements of the TM, particularly the handling of function words in paraphrasing models acquired from bilingual text. We also plan on constructing LMs that can evaluate speaker individuality in consideration of conversation context and experimenting on larger data from the web.

References

Abe M, Nakamura S, Shikano K, Kuwabara H (1988) Voice conversion through vector quantization. In: 1988 international conference on acoustics, speech, and signal processing, 1988. ICASSP-88, pp 655–658

Bannard C, Callison-Burch C (2005) Paraphrasing with bilingual parallel corpora. In: Proceedings of the 43rd annual meeting on association for computational linguistics, pp 597–604

Barzilay R, Lee L (2003) Learning to paraphrase: an unsupervised approach using multiple-sequence alignment. In: Proceedings of the 2003 conference of the North American chapter of the association for computational linguistics on human language technology, vol 1, pp 16–23. doi:10.3115/1073445.1073448. http://www.dx.doi.org/10.3115/1073445.1073448

Bond F, Isahara H, Fujita S, Uchimoto K, Kuribayashi T, Kanzaki K (2009) Enhancing the Japanese wordnet. In: Proceedings of the 7th workshop on Asian language resources, pp 1–8

Brill E, Moore RC (2000) An improved error model for noisy channel spelling correction. In: Proceedings of the 38th annual meeting on association for computational linguistics, pp 286–293

Chung C, Pennebaker JW (2007) The psychological functions of function words. In: Fiedler K (ed) Social communication. Psychology Press, New York, pp 343–359

Dagan I, Lee L, Pereira FC (1999) Similarity-based models of word cooccurrence probabilities. Mach Learn 34(1–3):43–69

Gosling SD, Rentfrow PJ, Swann WB Jr (2003) A very brief measure of the big-five personality domains. J Res Pers 37(6):504–528

Hiraoka T, Neubig G, Sakti S, Toda T, Nakamura S (2014) Construction and analysis of a persuasive dialogue corpus. In: 5th international workshop on spoken dialog systems (IWSDS)

Inui K, Fujita A (2004) A survey on paraphrase generation and recognition. J Nat Lang Process 11(5):151–198

Isard A, Brockmann C, Oberlander J (2006) Individuality and alignment in generated dialogues. In: Proceedings of the 4th international natural language generation conference, pp 25–32. http://www.dl.acm.org/citation.cfm?id=1706269.1706277

Isbister K, Nass C (2000) Consistency of personality in interactive characters: verbal cues, non-verbal cues, and user characteristics. Int J Hum Comput Stud 53(2):251–267. doi:10.1006/ijhc.2000.0368. http://www.dx.doi.org/10.1006/ijhc.2000.0368

Koehn P (2004) Statistical significance tests for machine translation evaluation. In: Conference on empirical methods on natural language processing, pp 388–395

Koehn P, Och FJ, Marcu D (2003) Statistical phrase-based translation. In: Proceedings of the 2003 conference of the North American chapter of the association for computational linguistics on human language technology - Volume 1, The 2013 conference of the North American chapter of the association for computational linguistics: human language technologies, pp 48–54. doi:10.3115/1073445.1073462. http://www.dx.doi.org/10.3115/1073445.1073462

Mairesse F, Walker MA (2011) Controlling user perceptions of linguistic style: trainable generation of personality traits. Comput Ling 37(3):455–488

Metze F, Englert R, Bub U, Burkhardt F, Stegmann J (2009) Getting closer: tailored human-computer speech dialog. Univ Access Inf Soc 8(2):97–108. doi:10.1007/s10209-008-0133-0. http://www.dx.doi.org/10.1007/s10209-008-0133-0

Miller GA (1995) Wordnet: a lexical database for English. Commun ACM 38:39–41

Neubig G, Nakata Y, Mori S (2011) Pointwise prediction for robust, adaptable Japanese morphological analysis. In: Proceedings of the 49th annual meeting of the association for computational linguistics: human language technologies: short papers, vol 2, pp 529–533. http://www.dl.acm.org/citation.cfm?id=2002736.2002841

Neubig G, Akita Y, Mori S, Kawahara T (2012) A monotonic statistical machine translation approach to speaking style transformation. Comput Speech Lang 26(5):349–370

Och FJ, Ney H (2002) Discriminative training and maximum entropy models for statistical machine translation. In: Proceedings of association for computational linguistics

Qian Y, Soong FK, Yan ZJ (2013) A unified trajectory tiling approach to high quality speech rendering. IEEE Trans Audio Speech Lang Process 21(2):280–290

Riesa J, Irvine A, Marcu D (2011) Feature-rich language-independent syntax-based alignment for statistical machine translation. In: Proceedings of the conference on empirical methods in natural language processing, pp 497–507

Takezawa T, Sumita E, Sugaya F, Yamamoto H, Yamamoto S (2002) Toward a broad-coverage bilingual corpus for speech translation of travel conversations in the real world. In: Proceedings of language resources and evaluation conference, pp 147–152

Teshigawara M, Kinsui S (2012) Modern Japanese "Role Language"(Yakuwarigo): fictionalised orality in Japanese literature and popular culture. Socioling Stud 5(1):37–58

Wang WY, Finkelstein S, Ogan A, Black AW, Cassell J (2012) Love ya, jerkface: using sparse log-linear models to build positive (and impolite) relationships with teens. In: Proceedings of the 13th annual meeting of the special interest group on discourse and dialogue, pp 20–29

Xu W, Ritter A, Dolan B, Grishman R, Cherry C (2012) Paraphrasing for style. In: Proceedings of computational linguistics 2012, pp 2899–2914. http://www.aclweb.org/anthology/C12-1177

Yamagishi J, Usabaev B, King S, Watts O, Dines J, Tian J, Guan Y, Hu R, Oura K, Wu YJ et al (2010) Thousands of voices for HMM-based speech synthesis—analysis and application of TTS systems built on various ASR corpora. IEEE Trans Audio Speech Lang Process 18(5): 984–1004

Chapter 14
A Study on Natural Expressive Speech: Automatic Memorable Spoken Quote Detection

Fajri Koto, Sakriani Sakti, Graham Neubig, Tomoki Toda, Mirna Adriani, and Satoshi Nakamura

Abstract This paper presents a study on natural expressive speech during public talks. Specifically, we focus on how people convey important messages that may be retained in the audience's consciousness. Our study aims to answer several questions. Why are some public speeches memorable and inspirational for the audience, while others are not? Why are some memorable/inspirational spoken quotes more popular than others? Being able to evaluate why certain spoken words are memorable/inspirational is not a trivial matter, and most studies on memorable quote detection are only limited to textual data. In this study, we use both linguistic and acoustic features of public speeches in TED talks. The results reveal that based on those linguistic and acoustic features, we are able to distinguish memorable spoken quotes and non-memorable spoken quotes with 70.4 % accuracy. Furthermore, we also analyze the important factors that affect the memorableness and popularity of spoken quotes.

Keywords Automatic quote detection • Memorable spoken quote • Popularity analysis

F. Koto
Nara Institute of Science and Technology, Ikoma, Nara, Japan
University of Indonesia, Kota Depok, Jawa Barat, Indonesia
e-mail: fajri91@ui.ac.id

S. Sakti (✉) • G. Neubig • T. Toda • S. Nakamura
Nara Institute of Science and Technology, Ikoma, Nara, Japan
e-mail: ssakti@is.naist.jp; neubig@is.naist.jp; tomoki@is.naist.jp; s-nakamura@is.naist.jp

M. Adriani
University of Indonesia, Kota Depok, Jawa Barat, Indonesia
e-mail: mirna@cs.ui.ac.id

G. Geunbae Lee et al. (eds.), *Natural Language Dialog Systems and Intelligent Assistants*, DOI 10.1007/978-3-319-19291-8_14

14.1 Introduction

Research related to spoken dialog systems has progressed from the traditional task-based frameworks to more sophisticated social agents (Dautenhahn 2007) that can engage the user and expressively convey the intended message. Consequently, understanding the ways humans express themselves and engage their listeners is becoming a more important factor in designing these sorts of systems. Here, we focus on studying natural expressiveness and its effects during public speeches.

Through history, the best speeches of all time normally feature memorable quotes that genuinely inspire the audience. For instance, the most famous quote of John F. Kennedy, "Ask not what your country can do, ask what you can do for your country," has inspired many generations since he gave this speech in January 1961.[1] More recent examples of inspirational public speech can be found on TED.[2] TED features talks of 5–25 min by skilled speakers on subjects including technology, entertainment, design. Many famous people have given speeches on TED and inspired people by their memorable words. Recently, TED has started "TED Quotes," which collects memorable quotes from TED talks, annotates them manually, groups them by category, and provides an easy way for people to share their favorite quotes. The most popular quotes can have more than a thousand shares.

While some public speeches may have inspired many individuals, they raise deeper questions. Why are some spoken words be memorable and inspirational, while some others are not? Why are some memorable quotes more popular than others? Answering these questions will be more challenging than just determining whether particular keywords appear in a given segment of speech as in spoken term detection research (Miller et al. 2007; Vergyri et al. 2006). Memorable quote detection involves the evaluation of what is being said by the speaker and how the audience reacts, even with or without particular keywords. The challenge lies in detecting generic pearls of wisdom expressed with unusual combinations of words.

We argue that there may be specific patterns or combination of words, as well as specific intonation or accent patterns which distinguish memorable spoken quotes from other spoken utterances. In this study, we attempt to answer these questions by developing a method for automatic detection of memorable spoken quotes and analyzing their popularity.

14.2 Memorable Spoken Quote Detection

Research related to memorable quote detection is still very limited. Bandersky et al. proposed an automatic detection of memorable quotes from books using linguistic features (Bandersky and Smith 2012). Research by Kolak et al. also proposed an

[1] http://www.ushistory.org/.

[2] http://www.ted.com/.

approach for automatically mining quotes from extremely large text corpora (Kolak and Schilit 2008). Similar work by Liang et al. automatically extracts quotations and allows for efficient retrieval of the semantically annotated quotes from news stories (Liang et al. 2010). Another study by Danescu-Niculescu-Mizil et al. attempted to investigate the effect of phrasing on a quote's memorability from movie scripts (Danescu-Niculescu-Mizil et al. 2012). While most techniques developed so far for memorable quote detection have focused primarily on the processing of text, we are interested in discovering memorable spoken quotes from natural speech.

14.2.1 Corpus Construction

To enable the system to learn to distinguish between memorable and non-memorable spoken quotes, we compiled a corpus from the TED website. The collected memorable quotes resulted in a total of 2118 speech transcription segment files. To construct a corpus for comparison, we also randomly selected a total of 2118 speech transcription segment files from the rest of the data and labeled them as non-memorable quotes.

Within TED, there is a "share" function that allows users to share their favorite quotes with others, and we utilize the number of shares as a measure of popularity. Here, we only focused on extreme cases and constructed a corpus with memorable quotes that have zero shares (labeled as non-popular quotes) and memorable quotes that have more than 50 shares (labeled as popular quotes). Here, all newly published quotes still have zero shares, and thus we exclude them from the data. In total, the corpus consists of 262 non-popular quotes and 179 popular quotes.

Further details of data construction can be found in our previous work (Koto et al. 2014).

14.2.2 Features of Spoken Quotes

Bandersky et al. defined three kinds of linguistic features useful for memorable quote detection: lexical, punctuation, and part-of-speech (POS) (Bandersky and Smith 2012). Following these linguistic features, we utilize lexical features (*#capital*, *#quantifier*, *#stops*, *beginStop*, *hasDialog*, *#abstract*) and POS (*#POS*, *hasComp*, *hasSuper*, *hasPP*, *#IGSeq[i]*) features. As we focus on spoken utterances of memorable quotes, punctuation features are excluded. In addition, we included *hasSynonym* and *hasAntonym* features in our experiment. Detailed descriptions of these features are shown in Table 14.1. For *#quantifier*, *#stop*, and *#abstract*

Table 14.1 Linguistic feature sets for a particular quote *s*

Feature	Description
Lexical	
#capital	Number of capitalized words in *s*
#quantifier	Number of universal quantifiers in *s*
#stops	Number of common stopwords in *s*
beginStop	True if *s* begins with a stopword, False otherwise
hasDialog	True if *s* contains at least one of say, says, said
#abstract	Number of abstract concepts (e.g., adventure, charity, stupidity) in *s*
Part of speech	
#POS	POS = noun, verb, adjective, adverb, pronoun
hasComp	True if *s* contains a comparative adjective or adverb, False otherwise
hasSuper	True if *s* contains a superlative adjective or adverb, False otherwise
hasPP	True if *s* contains a verb in past participle, False otherwise
hasSynonym	True if *s* contains two words that are synonymous, False otherwise
hasAntonym	True if *s* contains two words are antonyms of each other, False otherwise
#IGSeq[i]	Count of the POS sequence with *i*-th highest $IG(X, Y)$ [Eq. (14.1)] in *s*

features, we use 17 quantifiers,[3] 174 stop words,[4] and 176 abstract words,[5] respectively.

The *#IGSeq[i]* feature is used to analyze the pattern of POS sequences. Here, we generate feature of tri-POS sequences from the data, resulting in 5724 generated POS sequences. We then computed the information gain of all POS sequences in all memorable and non-memorable quotes based on Eqs. (14.1) and (14.2),

$$IG(X, Y) = H(X) - H(X|Y), \tag{14.1}$$

$$H(X) = -p(x)\log_2 p(x). \tag{14.2}$$

Feature *#IGSeq[i]* expresses the number of times the *i*-th POS sequence is contained in quote *s*, where *X* indicates the presence or absence of the POS sequence in current quote and *Y* indicates the type of quote (memorable or non-memorable). In this study, based on the information gain of all POS sequences, we selected only the top 250 of POS sequences as linguistic features.

While previous work has focused on lexical features, in this study we also include acoustic features. Specifically, we use the INTERSPEECH 2010 paralinguistic challenge configuration (IS10 Paraling features) (Schuller et al. 2010). It consists of 1582 features, which are obtained in three steps: (1) 38 low-level descriptors are extracted and smoothed by simple moving average low-pass filtering; (2) their first-order regression coefficients are added; (3) 21 functionals are applied. However, 16 zero-information features (e.g., minimum F0, which is always zero) are discarded.

[3]http://www.tesol-direct.com/guide-to-english-grammar/quantifiers.

[4]http://www.ranks.nl/resources/stopwords.html.

[5]http://www.englishbanana.com.

Finally, two single features for F0, number of onsets and turn duration, are added. More details of each feature can be found in Schuller et al. (2010) and Eyben et al. (2010).

14.2.3 *Classifier*

Based on this corpus, we develop a method for automatic detection of memorable spoken quotes. Specifically, we use both linguistic and acoustic features to distinguish between memorable quotes and non-memorable quotes of public speeches in TED talks. We investigated three classifiers: neural networks (NN) (Fu 1994), Naive Bayes (NB) (Cristianini and Taylor 2000), and support vector machines (SVM) (Lewis 1998). We also performed feature selection with forward algorithm approach to estimate the best feature set.

14.3 Experimental Set Up and Evaluation

14.3.1 *Set Up*

Linguistic features were extracted using NLTK (Bird 2006), while acoustic features were extracted using openSMILE toolkit[6] (Eyben et al. 2010). There are a total of 264 linguistic features and 1582 acoustic features. Here, we perform fivefold cross validation with 80 % of the corpus as training set, with the remainder of the corpus as the test set. Training of the prediction models was performed with Rapidminer[7] (Akthar and Hahne 2012).

14.3.2 *Memorable Quote Detection*

First, we conducted memorable quote detection for all features and classifiers (NN, NB, and SVM). Table 14.2 shows the performance of all classifiers after feature selection. As a comparison, we also include the performance of the classifier using the top 10 features of memorable quote detection proposed by Bandersky which are obtained by SVM weighting (denoted as "Baseline"). The results reveal that our proposed features give better accuracy than the baseline, and the best results were achieved by the use of acoustic features.

Next, we combine selected features from all classifiers into one union set of selected features. As there is some overlap of features, we finally have 12 linguistic features and 9 acoustic features in total. The result shows the accuracy of memorable

[6] http://www.opensmile.sourceforge.net/.

[7] http://www.rapidminer.com.

quote detection based on an SVM classifier, using: (1) 12 selected linguistic features only with 66.45 % accuracy, (2) nine selected acoustic features only with 68.06 %, and (3) combination of the selected linguistic and acoustic features with the highest, 70.4 % accuracy. The results reveal that the classifier with all features performs better than the classifier with linguistic or acoustic features only.

14.3.3 Memorableness and Popularity Analysis

We further analyze the features selected by the feature selection procedure. For acoustic features, the selected features are mainly F0, logMelFreqBand, and MFCC. By performing SVM weighting on these selected features, we found out that F0 had the highest weight. It indicates that the prosody of the utterance is a significant feature that distinguishes between memorable quotes and non-memorable quotes.

For linguistic features, the selected features include beginStop, #noun, #adjective, and some POS-tag sequences. The details of those POS-tag sequences including examples of word sequences are given in Table 14.3. CC-PRP-VBD is actually an amalgamation of two single sentences, a compound sentence. Based on Table 14.3, the sentences containing CC-PRP-VBD sequences tend to be non-memorable quotes. This indicates that memorable quotes seldom use conjunctions or they usually consist of single sentences. On the other hand, sentences with POS sequences of NN-VBZ-DT, JJ-NN-NN, PRP-VBZ-IN, and NN-VBZ-RB tend to be memorable quotes. These POS sequences are mainly used for definition, elaboration, and explanation types of sentences. Based on this data, we may argue

Table 14.2 Accuracy of memorable quote detection with fivefold cross validation for baseline and proposed features (the chance rate is 50.0 %)

	Baseline	Proposed	
Classifier	Linguistic (%)	Linguistic (%)	Acoustic (%)
Neural network	63.98	64.87	67.71
Naive Bayes	62.91	65.04	68.18
Support vector machine	64.80	66.71	68.08

Table 14.3 POS-tag sequences selected for memorableness analysis (*MQ* memorable quotes, *NM* non-memorable quotes)

Sequence	Example	#MQ	#NM
CC-PRP-VBD	But I thought, and I introduced	43	124
NN-VBZ-DT	Belief is the, education is a	155	45
JJ-NN-NN	National automobile slum, quiet screaming desperation	236	183
PRP-VBZ-IN	It is as, it is like	95	39
NN-VBZ-RB	Innovation is not, privacy is not	165	50

that memorable quotes tend to contain general statements about the world from the perspective of the speaker.

For the popularity analysis, the experiment was conducted utilizing only linguistic features, as people share their favorite quotes based only on text. Our highest classification result was achieved by naive Bayes with **69.40 %** accuracy. The accuracy of neural network and SVM are 68.48 % and 62.13 %, respectively.

14.4 Conclusion

In this study, we discussed the possibilities of automatically detecting the memorable spoken quotes in real public speeches based on linguistic and acoustic features. The results reveal that a classifier with both linguistic and acoustic features performs better than a classifier with linguistic or acoustic features only. By the use of this feature combination, we can distinguish between memorable quotes and non-memorable quotes with 70.4 % accuracy. Based on the analysis of the selected features, the results reveal that most memorable quotes have definition, elaboration, and explanation type sentences, and the prosody of utterances is a significant acoustic feature that distinguishes between memorable quotes and non-memorable quotes.

Acknowledgements Part of this work was supported by JSPS KAKENHI Grant Number 26870371.

References

Akthar F, Hahne C (2012) Rapid Miner 5 Operator Reference. Rapid-I GmbH. http://rapidminer.com/wpcontent/uploads/2013/10/RapidMiner_OperatorReference_en.pdf

Bandersky M, Smith, DA (2012) A dictionary of wisdom and wit: learning to extract quotable phrase. In: Proceedings of NAACHL-HLT, Montréal, Canada, pp 69–77

Bird S (2006) NLTK: the natural language toolkit. In: Proceedings of COLING/ACL on interactive presentation sessions, Sydney, pp 69–72

Cristianini N, Taylor JS (2000) An introduction to support vector machines and other Kernel-based learning methods. Cambridge University Press, Cambridge

Danescu-Niculescu-Mizil C, Cheng J, Kleinberg J, Lee L (2012) You had me at hello: how phrasing affects memorability. In: Proceedings of ACL, Jeju Island, pp 892–901

Dautenhahn K (2007) Socially intelligent robots: dimensions of human-robot interaction. Philos Trans R Soc B 362:679–704

Eyben F, Woeller M, Schuller B (2010) openSMILE—the Munich versatile and fast open-source audio feature extractor. In: Proceedings of multimedia (MM), pp 1459–1462

Fu L (1994) Neural network in computer intelligence. McGraw-Hill International Edition/MIT-Press, New York/Cambridge

Kolak O, Schilit BN (2008) Generating links by mining quotations. In: Proceedings of 9th ACM conference on hypertext and hypermedia, pp 117–126

Koto F, Sakti S, Neubig G, Toda T, Adriani M, Nakamura S. (2014) Memorable spoken quote corpora of TED public speaking. In: Proceedings of the 17th oriental COCOSDA, Phuket, pp 140–143

Miller DR, Kleber M, Kao CL, Kimball O, Colthurst T, Lowe SA, Gish H (2007) Rapid and accurate spoken term detection. In: Proceedings of INTERSPEECH, pp 314–317

Lewis DD (1998) Naive Bayes at forty: the independence assumption in information retrieval. In: Proceedings of ECML-98, Berlin, Heidelberg, pp 4–15

Liang J, Dhillon N, Koperski K (2010) A large-scale system for annotating and querying quotations in news feeds. In: Proceeding of 3rd international semantic search workshop, p 7

Schuller B, Steidl S, Batliner A, Burkhardt F, Devillers L, Muller CA, Narayanan SS (2010) The INTERSPEECH 2010 paralinguistic challenge. In: Proceedings of INTERSPEECH, Makuhari, pp 2794–2797

Vergyri D, Shafran I, Stolcke A, Gadde VRR, Akbacak M, Roark B, Wang W (2006) The SRI/OGI 2006 spoken term detection system. In: Proceedings of INTERSPEECH, pp 2393–2396

Chapter 15
Evaluation of a Fully Automatic Cooperative Persuasive Dialogue System

Takuya Hiraoka, Graham Neubig, Sakriani Sakti, Tomoki Toda, and Satoshi Nakamura

Abstract In this paper, we construct and evaluate a fully automated text-based cooperative persuasive dialogue system, which is able to persuade the user to take a specific action while maintaining user satisfaction. In our previous works, we created a dialogue management module for cooperative persuasive dialogue (Hiraoka et al., Reinforcement learning of cooperative persuasive dialogue policies using framing, Proceedings of international conference on computational linguistics (COLING), 2014), but only evaluated it in a wizard-of-Oz setting, as we did not have the capacity for natural language generation (NLG) and natural language understanding (NLU). In this work, the main technical contribution is the design of the NLU and the NLG modules which allows us to remove this bottleneck and create the first fully automatic cooperative persuasive dialogue system. Based on this system, we performed an evaluation with real users. Experimental results indicate that the learned policy is able to effectively persuade the users: the reward of the proposed model is much higher than baselines and almost the same as a dialogue manager controlled by a human. This tendency is almost the same as our previous evaluation using a wizard-of-Oz framework (Hiraoka et al., Reinforcement learning of cooperative persuasive dialogue policies using framing, Proceedings of international conference on computational linguistics (COLING), 2014), demonstrates that the proposed NLU and NLG modules are effective for cooperative persuasive dialogue.

Keywords Cooperative persuasive dialogue • Framing • Reinforcement learning • Dialogue modeling • Dialogue system

---

T. Hiraoka (✉) • G. Neubig • S. Sakti • T. Toda • S. Nakamura
Nara Institute of Science and Technology, Ikoma, Nara, Japan
e-mail: takuya-h@is.naist.jp; neubig@is.naist.jp; ssakti@is.naist.jp; tomoki@is.naist.jp; s-nakamura@is.naist.jp

G. Geunbae Lee et al. (eds.), *Natural Language Dialog Systems and Intelligent Assistants*, DOI 10.1007/978-3-319-19291-8_15

15.1 Introduction

There is ongoing research on applying reinforcement learning to persuasion and negotiation dialogues, which are different from traditional task-based dialogues (Georgila and Traum 2011; Georgila 2013; Paruchuri et al. 2009; Heeman 2009). In task-based dialogue, the system is required to perform the dialogue to achieve the user goal, eliciting some information from the user to provide an appropriate service. A reward corresponding to the achievement of the user's goal is given to the system. In contrast, in persuasive dialogue, the system convinces the user to take some action achieving a system goal, for example buying a particular product or agreeing to a particular plan (Georgila 2013). In previous work, we have proposed the paradigm of cooperative persuasive dialogue (Hiraoka et al. 2014b, 2013), where reward corresponding to the achievement of both the user's and the system's goal is given to the system. This paradigm is useful in situations where the user and the system have different, but not mutually exclusive, goals, an example of which being a sales situation where the user wants to find a product that matches their taste and the system wants to successfully sell a product, ideally one with a higher profit margin.

In previous reports, we have applied reinforcement learning to cooperative persuasive dialogue and evaluated the learned policy in a wizard-of-Oz setting (Hiraoka et al. 2014b). We modeled the cooperative dialogue based on partially observable Markov decision processes (POMDP), and system policies were learned with reinforcement learning. We introduced framing (Irwin et al. 2013), description of alternatives with emotionally charged words, as a system action. In this previous work, we evaluated the learnt policy by substituting a human wizard of Oz for natural language understanding (NLU) and the natural language generation modules (NLG). In this evaluation framework, the result of the evaluation is highly dependent on the ability of the human wizard, and the effect of NLU and NLG is discounted, potentially overstating the effectiveness of the system.

In this paper, we construct and evaluate the first fully automated text-based cooperative persuasive dialogue system. At first, we give a review of our previous research (Hiraoka et al. 2014a,b) about learning cooperative persuasive policies and then explain new modifications to the dialogue modeling, the newly implemented NLU and NLG models, and the evaluation. Experimental results indicate that the learned policy with framing is effective, even in a fully automatic system. The reward of the learnt policy with framing is much higher than baselines (a policy without framing and a random policy) and almost the same as a policy controlled by a human. This tendency is almost the same as the result of our previous research using the wizard-of-Oz framework (Hiraoka et al. 2014b).

15.2 Cooperative Persuasive Dialogue Corpus

In this section, we give a brief overview of cooperative persuasive dialogue and a human dialogue corpus that we use to construct the dialogue models and dialogue system described in later sections. In our collected persuasive dialogue corpus (Sect. 15.2.1), we define and quantify the actions of the cooperative persuader (Sect. 15.2.2). In addition, we annotate persuasive dialogue acts of the persuader from the point of view of framing (Sect. 15.2.3).

15.2.1 Persuasive Dialogue Corpus

The cooperative persuasive dialogue corpus (Hiraoka et al. 2014a) consists of dialogues between a salesperson (persuader) and customer (persuadee) as a typical example of persuasive dialogue. The salesperson attempts to convince the customer to purchase a particular product (decision) from a number of alternatives (decision candidates). More concretely, the corpus assumes a situation where the customer is in an appliance store looking for a camera, and the customer must decide which camera to purchase from five alternatives.

Prior to recording, the salesperson is given the description of the five cameras and instructed to try to convince the customer to purchase a specific camera (the persuasive target). In this corpus, the persuasive target is camera A, and this persuasive target is invariant over all subjects. The customer is also instructed to select one preferred camera from the catalog of the cameras,[1] and choose one aspect of the camera that is particularly important in making their decision (the determinant). During recording, the customer and the salesperson converse and refer to the information in the camera catalog as support for their dialogues. The customer can close the dialogue whenever they want, and choose to buy a camera, not buy a camera, or reserve their decision for a later date. The total number of dialogues is 34, and the total time is about 340 min.

15.2.2 Annotation of Persuader and Persuadee Goals

We define the cooperative persuader as a persuader who achieves both the persuader and persuadee goals and cooperative persuasive dialogue as a dialogue where both the persuader and persuadee goals have been achieved. To measure the salesperson's success as a cooperative persuader, we annotate each dialogue with scores corresponding to the achievement of the two participants' goals. As the

[1]The salesperson is not told this information about customer preferences.

Table 15.1 An example of positive framing

(Camera A is) able to achieve performance of comparable single-lens cameras and can fit in your pocket, this is a point.

persuader's goal, we use persuasive success measured by whether the persuadee's final decision (purchased camera) is the persuasive target or not. As the persuadees goal, we use the persuadee's subjective satisfaction as measured by results of a questionnaire filled out by the persuadee at the end of the dialogue (1: not satisfied; 3: neutral; 5: satisfied). Note that we assume a situation that is not a zero-sum game, and thus the persuader and persuadee goals are not mutually exclusive.

15.2.3 Annotation of Dialogue Acts

15.2.3.1 Framing

Framing is the use of emotionally charged words to explain particular alternatives and is known as an effective way of increasing persuasive power. The corpus contains tags of all instances of negative/positive framing (Irwin et al. 2013; Mazzotta and de Rosis 2006), with negative framing using negative words and positive framing using positive words.

The framing tags are defined as a tuple $\langle a, p, r \rangle$ where a represents the target alternative, p takes value NEG if the framing is negative and POS if the framing is positive, and r is a binary variable indicating whether or not the framing contains a reference to the determinant that the persuadee indicated was most important (for example, the performance or price of a camera). The user's preferred determinant is annotated based on the results of the pre-dialogue questionnaire.

Table 15.1 shows an example of positive framing (p =POS) about the performance of Camera A (a =A). In this example, the customer answered that his preference is the price of camera, and this utterance does not contain any description of price. Thus, r =NO is annotated.

15.2.3.2 General Purpose Functions (GPF)

The corpus also contains tags for traditional dialogue acts. As a tag set to represent traditional dialogue acts, we use the general-purpose functions (GPF) defined by the ISO international standard for dialogue act annotation (ISO24617-2 2010). All annotated GPF tags are defined to be one of the tags in this set.

15.3 Cooperative Persuasive Dialogue Modeling

The cooperative persuasive dialogue model proposed in our previous research (Hiraoka et al. 2014b) consists of a user-side dialogue model (Sect. 15.3.1) and a system-side model (Sect. 15.3.2).

15.3.1 User Simulator

The user simulator estimates two aspects of the conversation:

1. The user's dialogue acts
2. Whether the preferred determinant has been conveyed to the user (conveyed preferred determinant; CPD)

The user's dialogue acts are represented by using GPFs (e.g., question, answer, and inform). In our research, the user simulator chooses one GPF or *None* representing no response at each turn. CPD represents that the user has been convinced that the determinant in the persuader's framing satisfies the user's preference. For example, in Table 15.1, "performance" is contained in the salesperson's positive framing for camera A. If the persuadee is convinced that the decision candidate satisfies his/her preference based on this framing, we say that CPD has occurred (r=YES). In our research, the user simulator models CPD for each of the five cameras. This information is required to calculate reward described in Sect. 15.3.2. Specifically, GPF and CPD are used for calculating naturalness and persuasive success, which are elements of the reward function.

The user's GPF G_{user}^{t+1} and CPD C_{alt}^{t+1} at turn $t+1$ are calculated by the following probabilities:

$$P(G_{\text{user}}^{t+1} | G_{\text{user}}^{t}, F_{\text{sys}}^{t}, G_{\text{sys}}^{t}, U_{\text{eval}}), \tag{15.1}$$

$$P(C_{\text{alt}}^{t+1} | C_{\text{alt}}^{t}, F_{\text{sys}}^{t}, G_{\text{sys}}^{t}, U_{\text{eval}}). \tag{15.2}$$

G_{sys}^{t} represents the system GPF at time t and F_{sys}^{t} represents the system framing at t. These variables correspond to system actions, and are explained in Sect. 15.3.2. G_{user}^{t} represents the user's GPF at t, C_{alt}^{t} represents the CPD at t, and U_{eval} represents the users's original evaluation of the alternatives.[2] In our research, this is the camera selected by the user as preferred at the beginning of the dialogue. We use the persuasive dialogue corpus described in Sect. 15.2.1 for training the user simulator, considering the customer in the corpus as the user and the salesperson in the corpus as the system. We use logistic regression for learning Eqs. (15.1) and (15.2).

[2]Values of these variables are set at the beginning of dialogue and invariant over the dialogue.

Table 15.2 Features for calculating reward. These features are also used as the system belief state

$\mathrm{Sat}_{\mathrm{user}}$		$\mathrm{PS}_{\mathrm{sys}}$	Total time	N	System and user current GPF
	Frequency of system commissives		C_{alt} (for of 6 cameras)		System and user previous GPF
	Frequency of system question		U_{eval} (for of 6 cameras)		System framing

15.3.2 Dialogue Modeling: Learning Cooperative Persuasion Policies

For training the dialogue system using reinforcement learning, in addition to the user simulator, the reward, system actions, and belief state are required (Williams and Young 2007).

Reward is calculated using three factors: user satisfaction, system persuasive success, and naturalness. As described in Sect. 15.1, cooperative persuasive dialogue systems must perform dialogue to achieve both the system and user goals. Thus, reward at each turn t is calculated with the following equation:

$$r_t = (\mathrm{Sat}^t_{\mathrm{user}} + \mathrm{PS}^t_{\mathrm{sys}} + N^t)/3. \tag{15.3}$$

$\mathrm{Sat}^t_{\mathrm{user}}$ represents a five level score of the user's subjective satisfaction (1: not satisfied; 3: neutral; 5: satisfied) at turn t scaled into the range between 0 and 1. $\mathrm{PS}^t_{\mathrm{sys}}$ represents persuasive success (1: SUCCESS; 0: FAILURE) at turn t. N_t represents bi-gram likelihood of the dialogue between the system and user at turn t. Sat and PS are calculated with a predictive model constructed from the corpus described in Sect. 15.2.1 (Hiraoka et al. 2014a).

The **system action** $\langle G_{\mathrm{sys}}, F_{\mathrm{sys}}\rangle$ is a GPF/framing $\langle a, p\rangle$ pair representing the dialogue act of the salesperson. We construct a unigram model of the salesperson's dialogue acts $P(G_{\mathrm{sales}}, F_{\mathrm{sales}})$ from the original corpus, then exclude pairs for which the likelihood is below 0.005. As a result, we use the remaining 13 pairs as system actions.

The **belief state** is represented by the features used for reward calculation (Table 15.2) and the reward calculated at previous turn. Note that of the 8 features used for reward calculation, only C_{alt} cannot be directly observed from the system action or NLU results, and thus the system estimates it through the dialogue by using the following probability:

$$\sum_{\widehat{C^t_{\mathrm{alt}}}} P(\widehat{C^{t+1}_{\mathrm{alt}}} | \widehat{C^t_{\mathrm{alt}}}, F^t_{\mathrm{sys}}, G^t_{\mathrm{sys}}, U_{\mathrm{eval}}) P(\widehat{C^t_{\mathrm{alt}}}), \tag{15.4}$$

where $\widehat{C^{t+1}_{\mathrm{alt}}}$ represents the estimated CPD at $t+1$, $\widehat{C^t_{\mathrm{alt}}}$ represents the estimated CPD at t, and the other variables are the same as those in Eq. (15.2).

15.4 Modifications of the Cooperative Persuasive Dialogue Model

In this paper, we further propose two modifications to the cooperative dialogue models described in Sect. 15.3: (1) considering NLU recognition errors in the belief state, and (2) normalization of reward factors.

15.4.1 Considering NLU Recognition Errors

In the cooperative dialogue model in Sect. 15.3, we are not considering recognition errors of the NLU module. In previous research (Hiraoka et al. 2014b), we evaluated the policies based on the wizard of Oz, where a human was substituted for the NLU module, precluding the use of estimation methods used in ordinary POMDP-based dialogue systems (Williams and Young 2007). However, in this paper, we use a fully automatic NLU module, which might cause recognition errors, and thus some method for recovery is needed.

In this work, we modify the dialogue model to consider NLU recognition errors, incorporating estimation of the true user dialogue act (i.e., GPF) into the dialogue model. The estimation is performed according to the following equation:

$$P(G^{t+1}_{\text{user}}|H_{G_{\text{user}}}) = \frac{\sum_{G^{t}_{\text{user}}} P(H_{G^{t+1}_{\text{user}}}|G^{t+1}_{\text{user}})P(G^{t+1}_{\text{user}}|G^{t}_{\text{user}})P(G^{t}_{\text{user}})}{\sum_{G^{t+1}_{\text{user}}}\sum_{G^{t}_{\text{user}}} P(H_{G^{t+1}_{\text{user}}}|G^{t+1}_{\text{user}})P(G^{t+1}_{\text{user}}|G^{t}_{\text{user}})P(G^{t}_{\text{user}})}. \tag{15.5}$$

H_{user} represents the NLU result (described in Sect. 15.5.1) at t, and other variables are the same as those in Eqs. (15.1) and (15.2). $P(H_{G^{t+1}_{\text{user}}}|G^{t+1}_{\text{user}})$ represents a confusion matrix between the actual GPF and recognition result. To construct the confusion matrix, in Sect. 15.6.1, we perform an evaluation of NLU and use the confusion matrix from this evaluation for the estimation of Eq. (15.5). $P(G^{t+1}_{\text{user}}|G^{t}_{\text{user}})$ is calculated using maximum likelihood estimation over the persuasive dialogue corpus described in Sect. 15.2.1.

15.4.2 Normalization of the Reward Factors

The reward function in Sect. 15.3.2 considers three factors: persuasive success, user satisfaction, and naturalness. In the current phase of our research, we have no evidence that one of these factors is more important than the other for cooperative persuasive dialogue and thus would like to treat them as equally important. However, in Eq. (15.3) the scales (i.e., the standard deviation) of factors are different, and thus factors with a larger scale are considered as relatively important, and other factors are considered as relatively unimportant. For example, in our previous research

(Hiraoka et al. 2014b), the scale of naturalness N is smaller than other factors and as a result is largely ignored in the learning.

In this work, we fix this problem by equalizing the importance of reward factors through normalization with z-score. More concretely, the reward function of Eq. (15.3) is substituted with the following reward function:

$$r_t' = \frac{\text{Sat}_{\text{user}}^t - \overline{\text{Sat}_{\text{user}}^t}}{\text{Stddev}(\text{Sat}_{\text{user}})} + \frac{\text{PS}_{\text{sys}}^t - \overline{\text{PS}_{\text{sys}}}}{\text{Stddev}(\text{PS}_{\text{sys}})} + \frac{N^t - \overline{N}}{\text{Stddev}(N)}, \tag{15.6}$$

where variables with a bar represent the mean of variables without a bar, and the Stddev function represents the standard deviation of the argument. These statistics are calculated from simulated dialogue with the proposed dialogue model in the previous section, where actions are chosen randomly. We sampled the reward factor for 60,000 turns of the simulated dialogue (about 6000 dialogues) for calculating the statistics of each variable.

15.5 Text-Based Cooperative Persuasive Dialogue System

The main contribution of this paper is the construction of a fully automated text-based cooperative persuasive dialogue system. The structure of the system is shown in Fig. 15.1. In this section, we describe the construction of NLU (Sect. 15.5.1) and NLG (Sect. 15.5.2) modules that act as an interface between the policy module and the human user and are necessary for fully automatic dialogue.

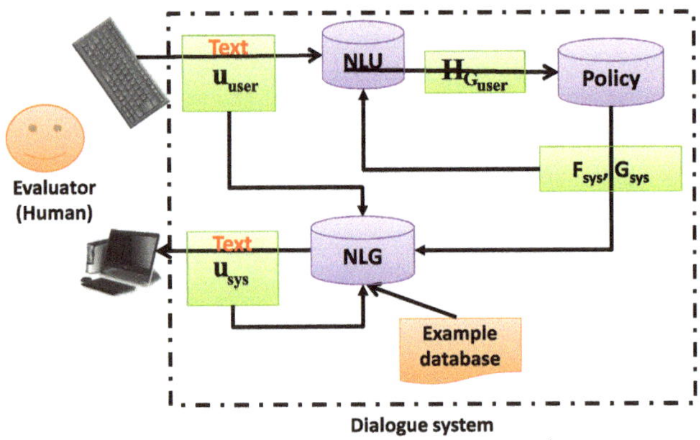

Fig. 15.1 Structure of our dialogue system. *Rectangles* represent information, and *cylinders* represent a system module

Table 15.3 Distribution of the GPF labels in the training data

Other	Question	SetQuestion	Propositional Question	Inform	Answer	Directive	Commissive
46	4	12	156	260	117	36	63

15.5.1 Natural Language Understanding

The NLU module detects the GPF in the user's text input u_{user} using a statistical classifier. In this paper, we use bagging, using decision trees as the weak classifier (Breiman 1996). We require the NLU to (1) be simple and (2) output the estimated classes with probability, and bagging with decision trees satisfies these requirements. The NLU uses many features (i.e., word frequency), and decision trees can select a small number of effective features, making a simple classifier. In addition, by using bagging, the confidence probability, which is determined by the voting rate of decision trees, can be attached to the classification result. We utilize Mark (2009) for constructing the bagging classifier.

As input to the classifier, we use features calculated from u_{user} and the history of system outputs (u_{sys}, $\langle G_{\text{sys}}, F_{\text{sys}} \rangle$). Features are mainly categorized into four types:

Uni: Unigram word frequency in the user's input
Bi: Bigram word frequency in the user's input
DAcl: The previous action of the system (i.e., GPF/framing pairs $\langle G_{\text{sys}}, F_{\text{sys}} \rangle$)
Unicl: Unigram word frequency in the previous system utterance

As we use Japanese as our target language, we perform morphological analysis using Mecab (Kudo et al. 2004) and use information about the normal form of the word and part of speech to identify the word.

As the NLU result $H_{G_{\text{user}}}$, eight types of GPF are output with membership probabilities. We use 694 customer utterances in the camera sales corpus (Sect. 15.2) as training data. In this training data, eight types of GPF labels are distributed as shown in Table 15.3.

15.5.2 Natural Language Generation

The NLG module outputs a system response u_{sys} based on the user's input u_{user}, the system's previous utterance u'_{sys}, and the system action $\langle G_{\text{sys}}, F_{\text{sys}} \rangle$. Though the dialogue assumed in this paper is focusing on a restricted situation, it is still not trivial to create system responses for various inputs. In order to avoid the large amount of engineering required for template-based NLG and allow for rapid prototyping, we decide to use the framework of example-based dialogue management (Lee et al. 2009).

Table 15.4 Part of the example database. The words surrounded by <> are inserted in correction

Speaker	Utterance	GPF	Framing
User	I want camera A. Do you have it? (私はAのカメラが欲しいんですけどありますか?)	PropQ	
Sys.	Yes, we do have <A camera in the store>. (<Aのカメラは店に>ありますよ)	Answer	
Sys.	What was the good point of camera A? (Aのカメラのどこがよかったんですか?)	Question	
User	Well, I like its shape, like a Monolith. (そうですね。このモノリスみたいな露骨な形が好だからです)	Answer	
Sys.	The main difference between camera A <and other cameras> is the sensor. (Aのカメラ<と他のカメラの大きな>違いはセンサーです) It is said that sensors are essential for a digital camera. (デジタルカメラはセンサーが命といわれています) The sensor of camera A is the same as that as a single-lens cameras. (Aのカメラのセンサーは一眼と同じセンサーを使ってるんですね。)	Inform	Pos A

We construct an example database $D = \{d_1, d_2, \ldots, d_M\}$ with M utterances by modifying the human persuasive dialogue corpus of Sect. 15.2. In the example database, the ith datum $d_i = \langle s, u, g, f, p \rangle$ consists of the speaker s, utterance u, GPF g, framing flag f, and previous datum p. In modifying the human persuasive dialogue corpus, we manually make the following corrections:

- Deletion of redundant words and sentences (e.g., fillers and restatements)
- Insertion of omitted words (e.g., subjects or objects) and sentences

Our example database consists of 2022 utterances (695 system utterances and 1327 user example utterances). An example of the database is shown in Table 15.4.

The NLG module determines the system response u_{sys} based on u_{user}, u'_{sys}, and $\langle G_{\text{sys}}, F_{\text{sys}} \rangle$. More concretely, our NLG modules performs the following procedure:

1. We define the response candidate set R according to whether there is user input ($u_{\text{user}} \neq \phi$) or not ($u_{\text{user}} = \phi$). If $u_{\text{user}} \neq \phi$, then we define R as the set of utterances r for which the previous utterance is a user utterance ($r.p.s =$ User). Conversely, if $u_{\text{user}} = \phi$, then we define R so $r.p.s =$ Sys.[3]

[3]In this paper, we use "." for representing the membership relation between variables. For example, Var1.Var2 means that Var2 is a member variable of Var1.

2. Response candidates R are scored based on the following similarity score:

$$\cos(r.p.u, u_{\text{input}}) = \frac{\text{words}(r.p.u) \cdot \text{words}(u_{\text{input}})}{|\ \text{words}(r.p.u)\ | \cdot |\ \text{words}(u_{\text{input}})\ |}, \tag{15.7}$$

$$u_{\text{input}} = \begin{cases} u'_{\text{sys}} & (u_{\text{user}} = \phi), \\ u_{\text{user}} & (u_{\text{user}} \neq \phi). \end{cases}$$

The cosine similarity cos between the previous utterance of the response sentence candidate $r.p.u(r \in R)$ and input sentence u_{input} is used for the scoring. u_{input} is set as u'_{sys} or u_{user} depending on u_{user}. The words function returns the frequency vector of the content words (i.e., nouns, verbs, and adjectives) weighted according to tf-idf.

3. The $r^*.u$ that has the highest score is selected as the output of the NLG module u_{sys}

$$r^* = \arg\max_{r \in R} \cos(r.p.u, u_{\text{input}}), \tag{15.8}$$

$$u_{\text{sys}} = r^*.u. \tag{15.9}$$

15.6 Experimental Results

In this section, we perform two forms of experimental evaluation. First, as a preliminary experiment, we evaluate the performance of the NLU module proposed in Sect. 15.5.1. Then, we evaluate the fully automatic persuasive dialogue system.

15.6.1 Evaluation for NLU Using Different Feature Sets

First, we evaluate the performance of the NLU module using different feature sets proposed in Sect. 15.5.1. We prepare four patterns of feature sets (Uni, Uni+DAcl, Uni+CAcl+Unicl, and Uni+CAcl+Bi) and evaluate the recognition accuracy of GPF labels in the customer's utterances. The evaluation is performed based on 15-fold cross-validation with 694 customer utterances described in Sect. 15.5.1.

From the experimental result (Fig. 15.2), we can see that NLU with Uni+CAcl+Bi achieves the highest accuracy, and thus we decided to use Uni+CAcl+Bi for NLU of the dialogue system in the next section. Focusing on the details of the misclassified GPFs, we show the confusion matrix for classification results of the NLU module with Uni+CAcl+Bi in Table 15.5. From this matrix, we can see that Answer is misclassified to Inform and that SetQ and Question

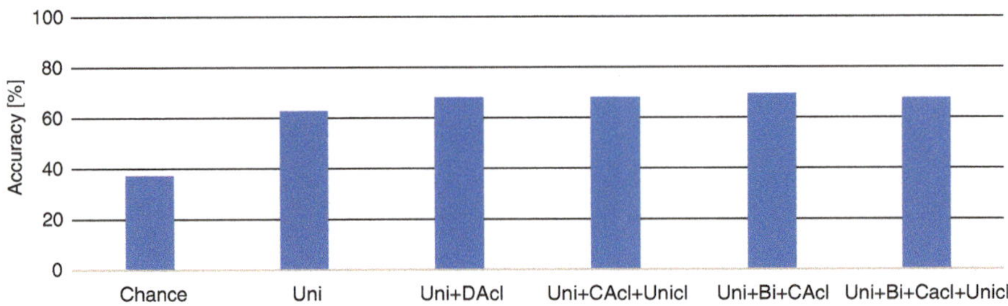

Fig. 15.2 Accuracy of the NLU module. The vertical axis represents accuracy and the horizontal axis represents the NLU feature set. Chance rate is an NLU module that always outputs inform

Table 15.5 The confusion matrix

Other	Commissive	PropQ	Directive	Answer	Inform	SetQ	Question	Classified as/true label
43	0	0	0	0	3	0	0	Other
6	31	2	4	0	20	0	0	Commissive
0	1	112	3	0	40	0	0	PropQ
2	2	6	13	0	13	0	0	Directive
0	3	5	0	53	56	0	0	Answer
1	12	4	4	9	230	0	0	Inform
0	0	10	0	0	2	0	0	SetQ
0	0	3	0	0	1	0	0	Question

Each row represents the distribution of the true GPF label. Each column represents the distribution of the NLU classification result

are misclassified into PropositionalQ. This result indicates that this module has difficulty in distinguishing dialogue acts in a hypernym/hyponym or sibling relationship.

15.6.2 Complete System Evaluation

In this section, we describe the results of the first user study evaluating fully automated cooperative persuasive dialogue systems. For evaluation, we prepare the following four policies.

Random: A baseline where the action is randomly output from all possible actions.

NoFraming: A baseline where the action is output based on the policy which is learned using only GPFs. For constructing the actions, we remove actions whose framing is not *None* from the actions described in Sect. 15.3.2. The policy is a greedy policy and selects the action with the highest score.

Framing: The proposed method where the action is output based on the policy learned with all actions described in Sect. 15.3.2 including framing. The policy is also a greedy policy.

Human: An oracle where the action is output based on human selection. In this research, the first author (who has no formal sales experience, but with experience of about 1 year in the analysis of camera sales dialogue) selects the action.

For learning the policies (i.e., NoFraming and Framing), we use Neural fitted Q Iteration (NFQ) (Riedmiller 2005). For applying NFQ, we use the Pybrain library (Schaul et al. 2010). The learning conditions follow the default Pybrain settings. We consider 3000 dialogues as one epoch and update the parameters of the neural network at each epoch. Learning is finished when the number of epochs reaches 20 (60,000 dialogues), and the policy with the highest average reward is used for evaluation.

We evaluate policies on the basis of average reward and correct response rate of dialogues with real users. The definition of the reward is described in Sect. 15.3.2, and the correct response rate is the ratio of correct system responses to all system responses. In the experiment, the dialogue system plays the salesperson, and the user plays the customer. At the end of the dialogue, to calculate the reward, the user answers the following questionnaire:

Satisfaction: The user's subjective satisfaction defined as a 5 level score of customer satisfaction (1: not satisfied; 3: neutral; 5: satisfied).

Final decision: The camera that the user finally wants to buy.

In addition, to calculate the correct response rate, we have the user annotate information regarding whether each system response is correct or not. Thirteen users perform one dialogue with the system obeying each policy (a total of four dialogues per user).

Experimental results for the reward are shown in Fig. 15.3. From these results, we can see that the reward of Framing is higher than that of NoFraming and Random and almost equal to Human. This indicates that learning a policy with framing is effective in a fully automatic text-based cooperative dialogue system. It is interesting to note that the tendency of those scores is almost the same as those of the wizard-of-Oz based experiment (Hiraoka et al. 2014b). The exception is that the naturalness of

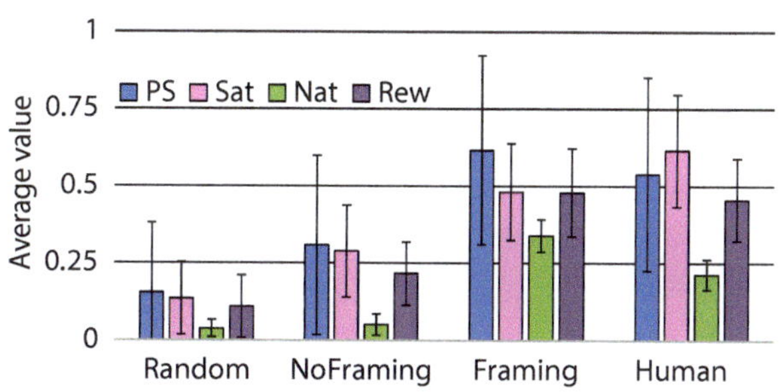

Fig. 15.3 Evaluation results for real users. Error bars represent 95 % confidence intervals. Rew represents the reward, Sat represents the user satisfaction, PS represents persuasive success, and Nat represents naturalness

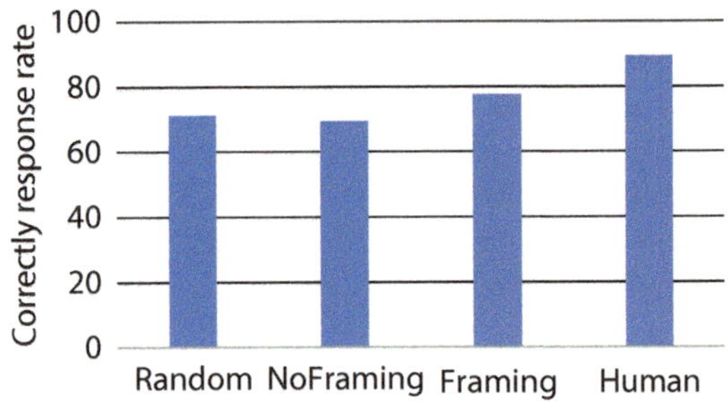

Fig. 15.4 Correct response rate of the system utterances

Framing in this experiment is higher than that of the wizard-of-Oz based experiment. Our hypothesis about the reason for this difference is that this is due to the effect of the modification of reward factors. In Sect. 15.4.2, we modified the importances of reward factors to be considered equally in learning the policy. Therefore, in the learning, naturalness is considered as an important factor, resulting in an increase of the naturalness score of Framing. It should be noted, however, that most of the subjects are different from the wizard-of-Oz based experiment we performed in previous work (Hiraoka et al. 2014b), and this might also affect the experimental result.

Experimental results for the correct response rate (Fig. 15.4) indicate that our cooperative persuasive dialogue system somewhat correctly responds to the user's input. The scores of all policies are higher than 70 %, and the score of Framing is about 77 %. In addition, even the Random policy achieves a score of about 70 %. One of the reasons for this is that NLG method used by our system (Sect. 15.5.2) is based on examples and thus is able to return natural responses that will only be judged as incorrect if they do not match the context.

15.7 Conclusion

In this paper, we presented a method for construction of a fully automatic cooperative persuasive dialogue system. Particularly, we focused on modifications to the policy learning and construction of NLU and NLG modules. We performed an evaluation of the constructed dialogue system with real users. Experimental results indicated that the proposed system is effective in text-based cooperative dialogue systems and that the tendency of each reward is almost the same as results of our previous research (Hiraoka et al. 2014b).

In the future, we plan to evaluate the system policies in more realistic situations that move beyond role-playing to real sales situations over more broad domains. We also plan to consider nonverbal information for estimating persuasive success and user satisfaction.

References

Breiman L (1996) Bagging predictors. Mach Learn 24(2):123–140

Georgila K (2013) Reinforcement learning of two-issue negotiation dialogue policies. In: Proceedings of the special interest group on discourse and dialogue (SIGDIAL)

Georgila K, Traum D (2011) Reinforcement learning of argumentation dialogue policies in negotiation. In: Proceedings of international speech (INTERSPEECH)

Heeman PA (2009) Representing the reinforcement learning state in a negotiation dialogue. In: Proceedings of IEEE automatic speech recognition and understanding workshop (ASRU)

Hiraoka T, Yamauchi Y, Neubig G, Sakti S, Toda T, Nakamura S (2013) Dialogue management for leading the conversation in persuasive dialogue systems. In: Proceedings of IEEE automatic speech recognition and understanding workshop (ASRU)

Hiraoka T, Neubig G, Sakti S, Toda T, Nakamura S (2014a) Construction and analysis of a persuasive dialogue corpus. In: Proceedings of the international workshop on spoken dialog systems (IWSDS)

Hiraoka T, Neubig G, Sakti S, Toda T, Nakamura S (2014b) Reinforcement learning of cooperative persuasive dialogue policies using framing. In: Proceedings of international conference on computational linguistics (COLING)

Irwin L, Schneider SL, Gaeth GJ (2013) All frames are not created equal: a typology and critical analysis of framing effects. Organ Behav Hum Decis Process 76(2):149–188

ISO24617-2: Language resource management-Semantic annotation frame work (SemAF). Part2: Dialogue acts. ISO (2010)

Kudo T, Yamamoto K, Matsumoto Y (2004) Applying conditional random fields to Japanese morphological analysis. In: Proceedings of conference on empirical methods in natural language processing (EMNLP), pp 707–710

Lee C, Jung S, Kim S, Lee GG (2009) Example-based dialog modeling for practical multi-domain dialog system. Speech Commun 51(5):466–484

Mark H, Eibe F, Geoffrey H, Bernhard P, Peter R, Ian HW (2009) The WEKA Data Mining Software: An Update; SIGKDD Explorations, 11(1)

Mazzotta I, de Rosis F (2006) Artifices for persuading to improve eating habits. In: AAAI spring symposium: argumentation for consumers of healthcare

Paruchuri P, Chakraborty N, Zivan R, Sycara K, Dudik M, Gordon G (2009) POMDP based negotiation modeling. In: Proceedings of the first MICON (modeling intercultural collaboration and negotiation), pp 66–78

Riedmiller M (2005) Neural fitted Q iteration - first experiences with a data efficient neural reinforcement learning method. In: Gama J, Camacho R, Brazdil PB, Jorge AM, Torgo L (eds) Machine learning: ECML. Springer, Berlin

Schaul T, Bayer J, Wierstra D, Sun Y, Felder M, Sehnke F, Ruckstiess T, Schmidhuber J (2010) Pybrain. J Mach Learn Res 11:743–746

Williams JD, Young S (2007) Partially observable Markov decision processes for spoken dialog systems. Comput Speech Lang 21(2):393–422

Chapter 16
Unknown Word Detection Based on Event-Related Brain Desynchronization Responses

Takafumi Sasakura, Sakriani Sakti, Graham Neubig, Tomoki Toda, and Satoshi Nakamura

Abstract The appearance of unknown words often disturbs communication. Most work on unknown words in spoken dialog systems deals with words that are uttered by the user, but which are not covered by the system's vocabulary. In this paper, we focus on detecting unknown words from the user side, in the case where the system utterance is unknown to the user. In particular, we develop a classifier based on Electroencephalography (EEG) signal from the user's brain waves, including the use of absolute power and Event-Related Desynchronization (ERD) features. The results show that we could detect the characteristics of brain waves at the time of unknown word perception significantly better than the chance rate.

Keywords Unknown word detection • EEG signal • Event-related desynchronization

16.1 Introduction

Skilled human communicators often adapt their language to suit the domain expertise of dialog partners. For example, a doctor will use technical medical terms when speaking to other medical professionals and simpler terms when speaking to patients. Therefore, it is desirable to develop dialog systems that can adapt to user conditions in a similar way. The challenge is to provide the ability to detect miscommunication and dynamically generate user-adaptive utterance variations.

One of major problems that causes miscommunication is the appearance of the unknown words. Various techniques in the spoken language understanding component have been proposed in order to detect and handle user words that are not covered by the system lexicon (Young 1994; Kai et al. 1998). On the other hand,

T. Sasakura • S. Sakti (✉) • G. Neubig • T. Toda • S. Nakamura
Nara Institute of Science and Technology, Ikoma, Nara, Japan
e-mail: sasakura.takafumi.sl4@is.naist.jp; ssakti@is.naist.jp, neubig@is.naist.jp, tomoki@is.naist.jp; s-nakamura@is.naist.jp

G. Geunbae Lee et al. (eds.), *Natural Language Dialog Systems and Intelligent Assistants*, DOI 10.1007/978-3-319-19291-8_16

methods to detect and handle the system utterances that might be unknown to the user have not been widely explored. Our study focuses on the latter issue. The aim is to detect when the user does not know one of the terms output by the system. Through this, the system may be able to estimate the knowledge level of the user and therefore have the capability to adapt and express the content in words that match the user's vocabulary (i.e., use a known synonym or describe the unknown word in other words).

However, having awareness of the user's state and detecting the user's (lexical) domain knowledge is not straightforward. One could imagine several ways to do so, including extracting paralinguistic information such as gaze and face expressions or performing explicit confirmations in which the user is queried about their understanding. However, these paralinguistic signals are not guaranteed to occur every time an unknown word occurs, and explicit confirmation is burdensome for the user. In this work we take a different approach, looking directly into the user's mind through electrophysiological measurements of brain waves. Specifically, we present a new way of detecting user misunderstanding in the form of unknown words based on the user's Electroencephalography (EEG) signal.

16.2 EEG Event-Related Desynchronization

EEG is an electrophysiological measurement of the brain activity at the human scalp surface whereby voltage variations of cortical field potentials are imaged (Regel 2009). It records electrical signals generated by the brain through electrodes placed on different points on the scalp and measures by comparing the voltage between two or more different sites. With regard to dialogue, Sridharan et al. (2012) presented NeuroDialog, which uses an EEG-based predictive model to detect system misrecognitions during live interaction. In this work, instead of system misunderstanding, we focus on detecting user misunderstanding in the form of unknown words.

When an unknown word is perceived, it is assumed that there is the matching process between the word and the memory. Sederberg et al. (2003) found that during memory encoding of later recalled nouns, power of the specific frequency band was significantly higher than for not recalled nouns. Sauseng et al. (2008) interpreted the result as memory matching between incoming visual information and stored (top-down) information. Klimesch (1999) presented EEG oscillations in the alpha and theta bands that reflect cognitive and memory performance in particular. In this study, we classify EEG data using the EEG state at the time of the perception of the unknown word and the change related to the event, which is called Event-Related Desynchronization (ERD) (Pfurtscheller and Aranibar 1977). The ERD value is expressed as the ratio of the decrease in band power of the target epoch (P_t) as compared to a reference interval (P_r), which is selected by experimenter before the target epoch, by using the simple equation:

$$\mathrm{ERD} = \frac{P_r - P_t}{P_r} \tag{16.1}$$

We extract ERD values as features from the EEG data. The mean of each feature is normalized to 0 and the standard deviation to 1. To improve performance, features are selected with the parameter subset selection forward algorithm (Hiruma et al. 2011), which is shown below.

1. Initialize each subset to consist of one feature.
2. Calculate the score J for each subset J given by

$$J = \frac{S_B}{S_W} \tag{16.2}$$

where

$$S_B = \sum_{j}^{L} N_j (m_j - M)^t (m_j - M) \tag{16.3}$$

$$S_W = \sum_{j}^{L} \sum_{i}^{N_j} (x_i - m_j)^t (x_i - m_j) \tag{16.4}$$

L is the number of classes, N is the number of subsets, and N_j is the number of subsets of class j. M is the means of all subset vectors, m_j is the mean of the subset vectors of class j, and x_i is a subset vector.
3. Select the features of the subset which had the highest score in the 2nd step.
4. Add a feature not included in the subset selected in 3rd step to the subset.
5. Repeat from second to 4th step until the maximum score falls.

As a baseline, we also test a system that uses only the power of each frequency band as features.

16.3 Experimental Set Up

16.3.1 Subjects and Stimuli Procedure

Six male Japanese-speaking subjects (23–24 years old in average) participated in the experiment. All participants were right-handed and had normal or corrected to normal vision. They don't have history of psychiatric or neurological illness or alcohol abuse, as well as no history of visual deficit. However, due to the effect of unnecessary components such as muscle artifacts, only five out of six EEG data were analyzed.

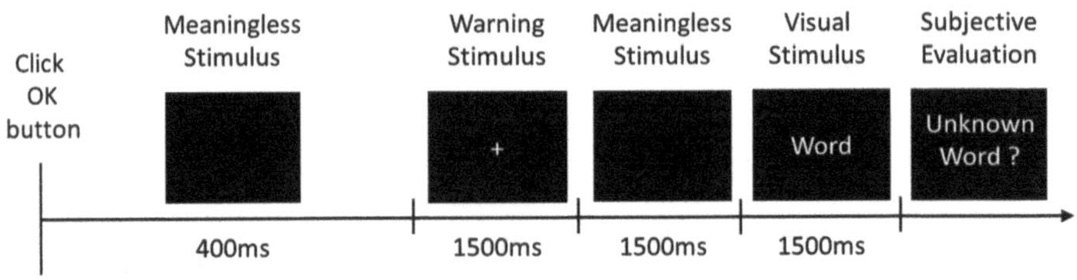

Fig. 16.1 An outline of the procedure of the experiment. When the values of ERD were calculated, the reference was during when the warning stimulus was presented, the target was during when the visual stimulus was presented

Three hundred Japanese noun words of 4 mora were presented to the subjects as visual stimuli. A mora is a Japanese subsyllabic unit which provides the root of rhythm (Otake et al. 1993). These words were constructed from Familiarity-controlled Word-lists 2007 Corpus (FW07) (Kondo et al. 2008). Originally, the list consists of words with four levels based on familiarity. In this study, we only used 150 words each from familiarity levels 1 and 4, which are the maximum and the minimum familiarity levels.

All subjects sat in a comfortable chair in a dark soundproofed room. The visual stimuli were presented on a 27 inch TV screen located 120 cm in front of them. The 300 words were presented visually at the center of the TV screen in white letters on a black background.

In this experiment, the subjects' task was to read a visually presented word and to answer a question about that stimulus. Figure 16.1 illustrates the stimuli procedure, which consists of following steps: (1) 4000 ms of meaningless stimulus; (2) 1500 ms of a plus mark "+" as a warning signal; (3) 1500 ms of another meaningless stimulus; (4) 1500 ms of a stimulus word (in Katakana) which was chosen at random; (5) subjective evaluation, where the subject gives a mark whether a shown word was unknown or known/overlooked. After the subjects completed the evaluation, they may click the OK button to start the next trial. These trials were repeated 300 times for all subjects.

16.3.2 EEG Recording

We recorded EEG from 29 sites on the scalp using a BrainAmp made by the Brain Product company. The grounding electrode was placed on both earlobes and the reference electrode to the apex of nose. To improve the signal to noise ratio, the impedance of each electrode was reduced to less than 5 kΩ using exclusive paste. EEG data was recorded with a sampling frequency of 1000 Hz.

We then cut the high frequency components such as muscle artifacts using a low-pass filter less than 40 Hz. Furthermore, trials including an amplitude more than 80 μV are excluded from the analysis. These procedures are called artifact reduction.

We extracted the target data, using EEG signals from successive 256 ms (256 points) time segments (windows or epochs) with 50 % overlap for 1024 ms from the time when visual stimuli were shown. A Hamming window was applied to each time segment to attenuate the leakage effect. Power density of the spectral components was then calculated based on a fast Fourier transformation (FFT). Furthermore, to calculate the power change (ERD), the same processing was carried out for the EEG data starting 1024 ms from when the warning stimuli were presented. Using these values, the ERD value was calculated.

16.3.3 SVM Classifier

The data was classified into two classes labeled according to the subjective evaluation results and the data of two classes was balanced. The feature used for classifier is selected from the power or the ERD value per each frequency band for seven time windows of each 29 channels. Because the EEG characteristics vary among individuals, we chose to make a separate classifier for each individual. After performing feature extraction for each subject, the selected features were used to train a classifier to distinguish between known or unknown words. Support vector machines (SVM) with the Radial Basis Function (RBF) Kernel were used for classification.

16.4 Results and Discussion

Figure 16.2 shows the means of the accuracy of tenfold cross-validation of five subjects. As a baseline, we use the chance rate. First, we apply a classifier simply on the absolute power of each frequency of EEG signals as features (denoted as "EEG-Power"). Second, we apply a classifier on ERD features (denoted as "EEG-ERD"). The results show that EEG-Power only provided a significant difference for three subjects. In contrast, the accuracy of classification using ERD ("EEG-ERD") was much higher than the chance rate in all subjects. Compared with EEG-Power, EEG-ERD provides a marginally significant difference over four subjects.

According to this result, we can see that it is important to capture the differential from the background signal. Because the accuracy of EEG-ERD is higher than EEG-Power in all subjects and there is marginally significant difference for most subjects, it is clear that the absolute value of power is not enough for prediction. Therefore, the user of differential features as in "EEG-ERD" provides a better solution.

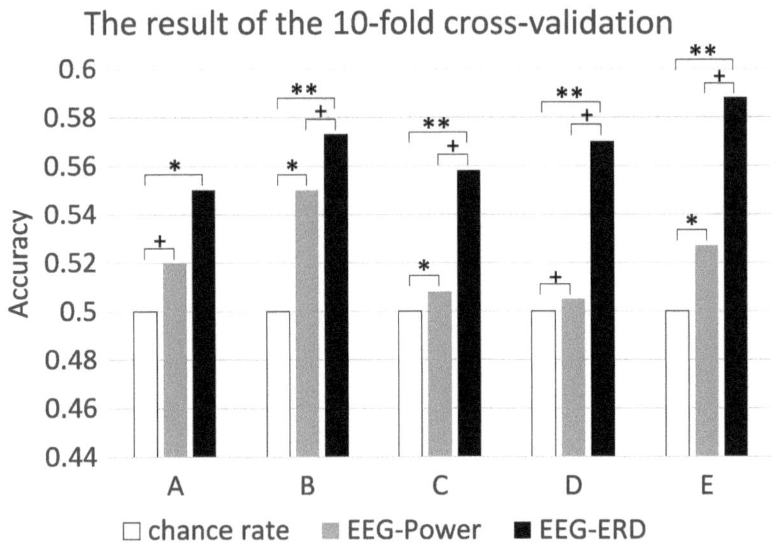

Fig. 16.2 Difference of the accuracy among the three kinds of features for the 5 subjects A-E. The bars marked *asterisk symbol* have a significant difference compared with the accuracy of the chance rate, the bar marked *plus symbol* has a marginally significant difference ($**p < 0.01$, $*p < 0.05$, $^{+}p < 0.10$, binomial test)

16.5 Conclusion

In this study, we detected unknown words from EEG when a subject perceived a word visually. As a result, both EEG-based classifiers (EEG-Power and EEG-ERD) showed a better performance than the chance rate. The best performance was obtained by classifier using ERD features (EEG-ERD).

Future work includes improvement of the performance of the classifier, experiments in an environment like a real conversation, and application to a multi-modal dialog system.

Acknowledgements Part of this work was supported by the Commissioned Research of National Institute of Information and Communications Technology (NICT) Japan, Microsoft CORE 10 Project, and JSPS KAKENHI Grant Number 26870371.

References

Hiruma N, Sagara K, Tanaka Y, Takeichi H, Yamashita O, Hasegawa R, Okabe T, Maeda T (2011) Brain communication : theory and application. IEICE Soc Conf 94(10):926

Kai A, Hirose Y, Nakagawa S (1998) Dealing with out-of-vocabulary words and speech disfluencies in an n-gram based speech understanding system. In: International conference on spoken language processing (ICSLP), vol 2, pp II–21

Klimesch W (1999) Eeg alpha and theta oscillations reflect cognitive and memory performance: a review and analysis. Brain Res Rev 29(2):169–195
Kondo T, Amano S, Sakamoto S, Suzuki Y (2008) Development of familiarity-controlled word-lists (fw07). IEICE Soc Conf Res Rep 107(432):43–48
Otake T, Hatano G, Cutler A, Mehler J (1993) Mora or syllable? speech segmentation in Japanese. J Mem Lang 32(2):258–278
Pfurtscheller G, Aranibar A (1977) Event-related cortical desynchronization detected by power measurements of scalp {EEG}. Electroencephalogr Clin Neurophysiol 42(6):817–826 doi:http://dx.doi.org/10.1016/0013-4694(77)90235-8. http://www.sciencedirect.com/science/article/pii/0013469477902358
Regel S (2009) The comprehension of figurative language: electrophysiological evidence on the processing of irony. Ph.D. thesis, Universitätsbibliothek
Sauseng P, Klimesch W, Gruber WR, Birbaumer N (2008) Cross-frequency phase synchronization: a brain mechanism of memory matching and attention. Neuroimage 40(1):308–317
Sederberg PB, Kahana MJ, Howard MW, Donner EJ, Madsen JR (2003) Theta and gamma oscillations during encoding predict subsequent recall. J Neurosci 23(34):10809–10814
Sridharan S, Chen YN, Chang KM, Rudnicky AI (2012) Neurodialog: an eeg-enabled spoken dialog interface. In: Proceedings of the 14th ACM international conference on multimodal interaction (ICMI '12), pp 65–66
Young SR (1994) Detecting misrecognitions and out-of-vocabulary words. In: IEEE international conference on acoustics, speech, and signal processing (ICASSP), vol 2, pp II–21

Chapter 17
An Analysis Towards Dialogue-Based Deception Detection

Yuiko Tsunomori, Graham Neubig, Sakriani Sakti, Tomoki Toda, and Satoshi Nakamura

Abstract When humans attempt to detect deception, they perform two actions: looking for telltale signs of deception and asking questions to attempt to unveil a deceptive conversational partner. There has been significant prior work on automatic deception detection that attempts to learn signs of deception. On the other hand, we focus on the second action, envisioning a dialogue systems that asks questions to attempt to catch a potential liar. In this paper, we describe the results of an initial analysis towards this goal, attempting to make clear which questions make the features of deception more salient. In order to do so, we collect a deceptive corpus in Japanese, our target language, perform an analysis of this corpus comparing with a similar English corpus, and perform an analysis of what kinds of questions result in a higher deception detection accuracy.

Keywords Automatic deception detection • Analysis of effective questions • Cross-lingual comparison

17.1 Introduction

Because it is known that it is not easy to detect deception during dialog, skilled interrogators use a number of techniques to detect deception (Ekman 1985), which include both looking for telltale signs and asking questions so that the features that give away a liar are more easily exposed (Vrij et al. 2011).

In recent years, there has been research on detecting deception automatically using machine learning techniques, and these works have achieved some degree of success. For example, Hirschberg et al. (2005) performed deception detection experiments on an English corpus including deception (the CSC corpus) using acoustic/prosodic and lexical features and achieved an accuracy (66.4 %) higher than the chance rate (60.2 %). In addition, Pérez-Rosas and Mihalcea (2014)

Y. Tsunomori • G. Neubig • S. Sakti (✉) • T. Toda • S. Nakamura
Nara Institute of Science and Technology, Ikoma, Nara, Japan
e-mail: tsunomori.yuiko.tq1@is.naist.jp; neubig@is.naist.jp; ssakti@is.naist.jp; tomoki@is.naist.jp; s-nakamura@is.naist.jp

G. Geunbae Lee et al. (eds.), *Natural Language Dialog Systems and Intelligent Assistants*, DOI 10.1007/978-3-319-19291-8_17

reported that there are differences in the lexical characteristics of deception between cultures or languages, although they make no mention of acoustic/prosodic features.

It should be noted that this previous research deals with only detecting deception in a particular already performed dialogue. In the analogy to human interrogators, this is equivalent to "looking for the telltale signs of deception," which, while important, is only half of the interrogators job. The other half, asking questions to cause deception features to be exposed, has not been covered in previous work. In our work, we envision a deception detecting dialogue system that can detect deception by not only looking for the telltale signs but also by asking questions to cause features of deception to be exposed. In this paper, we take a first step towards this goal by identifying not only which features can be used to create a deception detecting classifier but also which types of questions can cause a deceiver to show signs of deception. If these questions are made clear, in future work it will be possible to create a dialogue system that focuses on these questions and thus may be more effective at eliciting signs of deception.

In this research, we make two contributions. The first is that, as our target language is Japanese, we collect a Japanese corpus modeled after similar English corpora of detective speech. We perform deception detection experiments using these corpora and comparing features, both lexical and acoustic/prosodic, that can be used to detect deception effectively in Japanese and English. The second contribution is that we analyze which types of questions made by the interrogator make it possible to detect deception effectively on this corpus. Specifically, we examine the dialog act of questions that elicit utterances that are easy or difficult to classify.

17.2 Collection and Annotation of the Japanese Deception Corpus

Before performing research on data-driven deception detection, it is necessary to have a corpus, and a number of resources have been created in previous works. The CSC corpus (Hirschberg et al. 2005) recorded interviews where subjects were encouraged to lie to an interviewer and were motivated by financial incentive to deceive successfully. Interviews were performed in English, with a total of 22 interviews ranging from 25 to 50 min. Furthermore, there is the Idiap Wolf Corpus (Hung and Chittaranjan 2010), an audio-visual corpus containing natural conversational data of volunteers who took part in a competitive role-playing game in a group conversational scenario. Four groups of 8–12 people were recorded in English.

However, while excellent resources exist for English, there are fewer resources for other languages. In Japanese, there is the Indian Poker corpus (Ohmoto et al. 2009), an audio-visual corpus containing natural conversational data of 18 subjects who took part in 3-person games of Indian poker. However this resource is not publicly available, and because we assume a one-on-one dialogue system, a corpus recorded with three participants is not suitable for our research. Thus, as a first step in our research, we collect a corpus of deceptive utterances with two goals: first to allow comparative studies of deception detection in speech across languages and cultures and also to provide further resources for our work on deception detecting dialogue systems, which will use Japanese as its target language. To do so, we collect dialogs, make transcriptions, and annotate "lie" labels under the same conditions as the CSC corpus (Hirschberg et al. 2005).

17.2.1 Corpus Collection

In order to collect our corpus of deceptive speech, we follow the recording paradigm of the CSC corpus, as we describe below. As an example of scenes in which deception regularly occurs, the dialog recording assumes a self-presentational dialogue (DePaulo et al. 2003) between an interviewer and interviewee. The recording process is as follows:

1. The experimenter tells subjects that the experiment seeks to identify individuals who fit a "target profile" for six areas (politics, music, geography, food, interactive, and survival).
2. The subjects take a written test in the six areas before starting the interview.
3. The test scores are manipulated so that all subjects score too high to fit the profile in two areas, too low in 2, and correctly in 2. The experimenter tells the subjects the score.
4. The subjects are told that the experiment is actually about identifying people who can convince others that they fit the target profile in all areas. They are told that those who succeeded at deceiving the interviewer into believing that they fit the target profile in all areas can get a prize.
5. The subjects attempt to convince the interviewer that their scores in each of the six areas matched the target profile. The interviewers' task is determining how subjects had actually performed, and the interviewer is allowed to ask any questions other than those that were actually part of the tasks the interviewee had performed.

Two people were recruited as interviewers, and ten people were recruited as subjects. The total number of dialogs is 10, and the total time is about 150 min. The total number of utterances is 1069 and the total number of sentence-like units (SUs) is 1671, where an SU is a unit that divides utterances by punctuation marks and stops. We have named this corpus the "Japanese Deception Corpus (JDC)" and make it available for research purposes.[1] We show part of the JDC corpus in Table 17.1.

[1] http://ahclab.naist.jp/resource/ja-deception/.

Table 17.1 Example dialog (I/:interviewer, P/:subject)

Speaker	Transcription	Label
I	音楽に関して,あなたはマッチしていましたか？ How did you do on the music section?	
P	はい,マッチしていました. I matched the desired profile on that section.	Lie
I	それはなぜだと思いますか？ Why do you think so?	
P	えーと,そこそこ答えれたからです. Uh, I was able to answer so-so.	Truth
P	小さい頃からずっとピアノをやっていたので. I have played piano since I was a child.	Lie

17.2.2 *Annotation*

In order to label the veracity of subjects' SUs, we asked all subjects to push a "truth" or "lie" button during the interview for each SU. SUs including a lie in any parts are defined as a lie. Labels for lies were obtained automatically from button-push data and hand-corrected for alignment. The number of SUs labeled "truth" was 1401 and the number labeled "lie" was 270.

17.3 Features for Deception Detection

In order to perform deception detection experiments, it is necessary to define features that may be indicative of deception. Based on previous research (Hirschberg et al. 2005), we extract lexical and acoustic/prosodic features which may characterize deceptive speech. The extracted features are reported in Table 17.2.

- **Acoustic/prosodic features**
  As acoustic/prosodic features, we use fundamental frequency F_0, power, and phoneme duration. F_0 is obtained using the Snack Sound Toolkit (Sjolander 2004), and phoneme duration is obtained using Kaldi (Povey et al. 2011).
- **Lexical features**
  To extract lexical features, we first perform word segmentation and POS tagging of the Japanese sentences using MeCab (Kudo et al. 2004) and then use this information to calculate features. Of the listed features, topic indicates the test area under consideration, and noise indicates the presence of a cough or a sound resulting from the subject contacting the microphone. The frequency of positive-emotion words is extracted using Semantic Orientations of Words (Takamura et al. 2005). In addition to the previously proposed features for English, we add "the Japanese particles at the end of the sentence" which takes advantage of the

Table 17.2 Acoustic/prosodic, lexical, and subject-dependent features

Category	Description
Lexical	Topic, laugh, noise, disfluency, third person pronoun, denial, yes/no, end of sentence, verb base form, cue phrase, question, positive words, agree, filled pause
F_0	Median, percentage of median in SU
Phoneme duration	Vowel, average, max
Power	Average, First and last frame of SU
Subject-dependent	Gender, frequency of filled pause and cue phrase

fact that sentence final particles indicate when the speaker has confidence in their utterance (the "yo" particle), is attempting to seek the agreement of the listener (the "ne" particle) or other similar factors.

- **Subject-dependent features**
  We also extract features related to the characteristics of the subject. We use the gender, the frequency of cue phrases (e.g., well, actually, basically), and the frequency of filled pauses.

17.4 Deception Detection Experiments

Based on the data and features described in the previous sections, we first perform experiments on binary classification between deceptive and nondeceptive utterances. To solve this classification, we use Bagging of decision trees (Breiman 1996) as implemented in the Weka toolkit, which gave the best performance of the methods that we tested. The evaluation of the experiments is performed by leave-one-out cross-validation which uses 1670 SUs for training and 1 SU for testing.

17.4.1 Discussion

Table 17.3 shows the classification results. "Japanese" is the classification rate using the JDC corpus that we described in Sect. 17.2, and "English" is the classification rate using the CSC corpus. "Human" indicates the accuracy of manual classification, where utterances are classified by a different person from the subjects. He classified each SU without considering the context.

In Japanese, the accuracy of the classification using acoustic/prosodic and subject-dependent features is the highest, higher than the chance rate by about 7 %. Similarly in English, the accuracy using acoustic/prosodic and subject-dependent features is also highest, higher than the chance rate by about 17 %. The accuracy of utterances classified by humans is mostly the same as the chance rate, demonstrating the difficulty of deception detection for humans. As well, the rate using lexical

Table 17.3 Classification accuracy and deception detection F-measure for acoustic/prosodic (AP), lexical (L) and subject-dependent (S) features

Features	Japanese		English	
	Rate (%)	F-measure (%)	Rate (%)	F-measure (%)
Chance rate	83.8	0.0	71.4	0.0
AP	90.5	60.2	86.8	74.5
L	84.2	7.6	71.4	14.7
AP+S	90.7	61.4	88.1	77.7
L+S	85.2	31.5	76.8	52.9
AP+L	89.9	56.9	86.8	74.6
AP+L+S	90.2	58.1	87.8	77.2
Human	83.0	28.4		

Table 17.4 Accuracy between subjects (AP+L+S)

Subject	A	B	C	D	E	F	G	H	I	J
Chance rate (%)	93.5	89.4	92.4	76.9	78.7	75.9	64.9	82.9	73.1	83.9
Accuracy (%)	93.2	89.4	93.2	80.2	86.5	88.6	84.7	87.4	73.1	93.7

Table 17.5 Example dialogs (G/:higher accuracy, A/:lower accuracy)

Subject	Transcription
G	SNはー，そうですね，まぁMP七八割は答えれたかなっていう位ですね． SN Ah, so uh MP, may be I was able to answer for 70 % of the test.
A	たぶん大丈夫だと思います． I think that it's probably OK.

features alone in Japanese and English is almost equal to the chance rate. Because the accuracy of classification adding the frequency of cue phrases and filled pauses improved over this, we can see that subject-dependent features are effective to detect deception in both English and Japanese. Finally, we measured statistical significance between the results using Fisher's exact test and found significant differences between the chance rate and systems using acoustic/prosodic features, acoustic/prosodic + subject-dependent, acoustic/prosodic + lexical, and acoustic/prosodic + lexical + subject-dependent ($p < 0.01$).

Table 17.4 shows the accuracy of deception detection among subjects. Additionally, Table 17.5 shows the example dialogs between subjects in which deception detection is easy and difficult. SN is a noise and MP is a disfluency. The examples are subjects' replies to the interviewers' questions about the result of the test, and subject G is a speaker with a high deception detection accuracy and A is a speaker who has a low deception detection accuracy. It can be seen that A has many SN and MP, with an unsteady voice. On the other hand, G doesn't have many distinguishing differences from true utterances.

Table 17.6 Effective features

Category	English	Japanese
Lexical	Noise, third person pronoun, YesNo	Verb base
Subject-dependent	Frequency of cue phrase	
F_0	Median	Median
Phoneme duration	Average, vowel	Vowel
Power	Average, first and last frame of SU	Last frame of SU

17.4.2 Cross-Lingual Comparison of Effective Features

In this section we compare the difference in features that are effective in deception detection across the two languages. Using best-first search, we did feature selection maximizing the rate of classification on the training data. Table 17.6 shows the resulting selected features. As acoustic/prosodic features, the median of F_0, average of vowel phoneme duration, and the last frame of power were found effective for both Japanese and English. Potential reasons why these features were effective for both Japanese and English are as follows:

- Last frame of power
  Generally, the change of feelings (such as uncertainty) tend to appear at the end of an utterance.
- Median of F_0
  It is said that people often change voice tone when they tell a lie (DePaulo et al. 2003).
- Vowel duration
  It is possible that people tend to speak at different speeds when lying or telling the truth.

As lexical features, the result were greatly different Japanese and English. In English, noise, third person pronoun, and containing "Yes" or "No" were effective. On the other hand, the lexical features used in this research were largely ineffective, with only containing a verb base form proving effective in Japanese.

17.5 Analysis of Types of Questions That Detect Deception Effectively

As our final goal is to build a dialogue system that can perform or aid deception detection, in this section we analyze what kind of questions this system can perform to make it easier to detect deception. Assume that we have question of the interviewer q and its corresponding response r. In order to perform this analysis, we separate all responses r into classes based on some feature of the corresponding q, then measure the accuracy of deception detection for each class. If a particular class has higher deception detection accuracy, it can be said that q of this type are effective at drawing out features of deception that can be easily detected automatically and are thus questions that a deception detecting dialogue system should be focusing on.

17.5.1 Analysis of Question Dialogue Act

While this is a general framework for analyzing which kinds of questions are effective for deception detection, in this paper we specifically hypothesize that the variety (dialogue act) of the interviewers' utterance has an effect on the ease of detecting deception. In order to test this hypothesis, we use the dialogue act of q as the class into which we divide the responses r. Each utterance of the interviewer is annotated with a general-purpose function (GPF) defined by the ISO international standard for dialog act annotation (ISO24617-2, 2010). In this paper, annotators assign GPF manually. For approximately 10 % of the corpus, two annotators annotate GPFs and the mean rate of agreement is 80 %. Of these, we focus on situations where the annotator performs one of the following dialogue acts. The definitions of each dialogue act are quoted from the standard:

- **CheckQ**
  Communicative function of a dialogue act performed by the sender, S, in order to know whether a given proposition is true, about which S holds an uncertain belief that it is true. S assumes that addressee A knows whether the proposition is true or not and puts pressure on A to provide this information.
- **ChoiceQ**
  Communicative function of a dialogue act performed by the sender, S, in order to know which one from a given list of alternative propositions is true; S believes that exactly one element of that list is true; S assumes that the addressee, A, knows which of the alternative propositions is true, and S puts pressure on A to provide this information.
- **ProQ**
  Communicative function of a dialogue act performed by the sender, S, in order to know whether a given proposition is true. S assumes that A knows whether the proposition is true or not and puts pressure on A to provide this information.[2]
- **SetQ**
  Communicative function of a dialogue act performed by the sender, S, in order to know which elements of a certain set have a named property. S puts pressure on the addressee, A, to provide this information. S believes that at least one element of the set has the named property, and S assumes that A knows which are the elements of the set that have the property.

[2]This is a superset of checkQ, so ProQ in this work indicates all ProQ that are not CheckQ.

Table 17.7 Detail of the classification results

	Recall(%)		Rate(%)	
Class	True	Lie	Accuracy	Chance rate
CheckQ	99.4	83.3	95.3	77.7
ChoiceQ	100.0	80.0	96.4	78.6
ProQ	100	69.1	91.5	66.7
SetQ	99.5	73.7	94.0	75.8

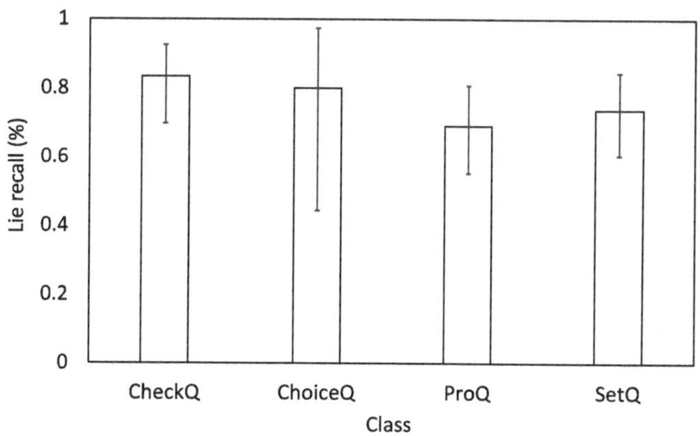

Fig. 17.1 Lie recall corresponding to each question

In Table 17.7, we show classification results for the subjects' SUs corresponding to each type of labeled GPF in the interviewers' utterance. In this case, we are most interested in the case where lies are correctly classified as lies (lie recall), as these indicate the possibility that the system can detect when the conversational partner is lying. In Fig. 17.1, we show lie recall. Confidence interval $p < 0.05$ is calculated by Clopper–Pearson method.

From these results, we can see that the category with the highest rate of lies that are correctly classified as lies is for SUs corresponding to CheckQ. In responses to CheckQ questions, subjects tend to talk about the previous speech again when interviewer asks them to confirm previous information. This is interesting in that Meyer (2011) reported that interviewers often let liars talk about same speech to detect deception. The result that CheckQ is most effective to detect deception is in concert with this observation. On the other hand, the lowest rate of lie classified as lies is SUs corresponding to ProQ, which conceivably put less pressure on the interviewee, as they only need to answer yes or no.

In addition, in Table 17.8 we show the number of words per SU in the interviewee response to each question. From this table, we can see that the length of utterances corresponding to CheckQ is the shortest. Again, this is in concert with the observation of Meyer (2011), lies are more easily exposed from the extremely short utterances.

Table 17.8 Mean length of SUs corresponding to each question

CheckQ	ChoiceQ	ProQ	SetQ	Average
6.4	12.2	11.8	18.1	12.3

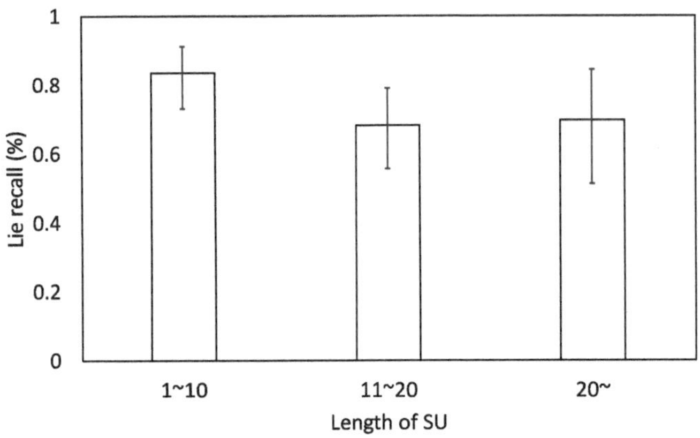

Fig. 17.2 Lie recall corresponding to each question length

17.5.2 Analysis of Question Length

It is important to ask questions to cause deception features to be exposed to detect deception. By asking about the details of the story and shaking the confidence of the partner, we can cause deception features to be exposed (Meyer 2011). Particularly, for a question to let you make the same talk again such as CheckQ, subjects cannot help answering the small point of the made-up story. In this case, we think that the question length is important because subjects think about a made-up story while a question. We hypothesize that deception features are exposed easily when interviewer asks the shorter question, the subjects can lie skillfully when interviewer asks the longer question.

To assess this hypothesis, we search the detection rate corresponding to each question length. In Fig. 17.2, we show lie recall corresponding to each question length. Confidence interval $p < 0.05$ is calculated by Clopper–Pearson method. For example, $1 \sim 10$ is the lie recall that classified for subjects' SUs corresponding to $1 \sim 10$ words question. Because lie recall corresponding to $1 \sim 10$ words is the highest, the short question is effective to detect deception. We think that the subjects' lies are exposed easily by asking short questions. We will analyze answer corresponding to each question length.

17.6 Conclusion and Future Work

In this paper, we described the collection of a Japanese deception corpus and experiments in detecting deception. We performed classification using features that were shown to be effective in English by previous research (Hirschberg et al. 2005) and confirmed that these features were also effective to some extent for Japanese. We also performed an analysis of the relationship between types of questions an interviewer makes and the ease of detecting deception. We confirmed that Check questions were the most effective variety to elicit utterances that make it easier to detect deception. We confirmed that the deception features are exposed easily by asking short questions.

In future research, we plan to further analyze other aspects of questions that may influence the accuracy of deception detection. We will also perform the actual implementation of the deception detecting dialogue system based on our analysis of these effective questions.

References

Breiman L (1996) Bagging predictors. Mach Learn 24(2):123–140

DePaulo BM, Lindsay JJ, Malone BE, Muhlenbruck L, Charlton K, Cooper H (2003) Cues to deception. Psychol Bull 129(1):74

Ekman P (1985) Telling lies. W. W. Norton & Company, New York

Hirschberg JB, Benus S, Brenier JM, Enos F, Friedman S, Gilman S, Girand C, Graciarena M, Kathol A, Michaelis L, Pellom B, Shriberg E, Stolcke A (2005) Distinguishing deceptive from non-deceptive speech. In: Proceedings of Eurospeech, Losbon

Hung H, Chittaranjan G (2010) The IDIAP wolf corpus: exploring group behaviour in a competitive role-playing game. In: Proceedings of the international conference on multimedia. ACM, New York, pp 879–882

Kudo T, Yamamoto K, Matsumoto Y (2004) Applying conditional random fields to Japanese morphological analysis. Proc EMNLP 4:230–237

Meyer P (2011) Lie spotting. Griffin, New York

Ohmoto Y, Ueda K, Ohno T (2009) A method to detect lies in free communication using diverse nonverbal information: towards an attentive agent. In: Active media technology. Springer, Berlin, pp 42–53

Pérez-Rosas V, Mihalcea R (2014) Cross-cultural deception detection. In: Proceedings of association for computational linguistics (ACL), pp 440–445. http://aclweb.org/anthology/P14-2072

Povey D, Ghoshal A, Boulianne G, Burget L, Glembek O, Goel N, Hannemann M, Motlicek P, Qian Y, Schwarz P, Silovsky J, Stemmer G, Vesely K (2011) The Kaldi speech recognition toolkit. In: Proceedings of ASRU

Sjolander K (2004) Tcl/tk snack toolkit http://www.speech.kth.se/snack/

Takamura H, Inui T, Okumura M (2005) Extracting semantic orientations of words using spin model. Proceedings of association for computational linguistics (ACL), pp. 133–140

Vrij A, Granhag PA, Mann S, Leal S (2011) Outsmarting the liars: toward a cognitive lie detection approach. Curr Dir Psychol Sci 20(1):28–32

Chapter 18
Pair Me Up: A Web Framework for Crowd-Sourced Spoken Dialogue Collection

Ramesh Manuvinakurike and David DeVault

Abstract We describe and analyze a new web-based spoken dialogue data collection framework. The framework enables the capture of conversational speech from two remote users who converse with each other and play a dialogue game entirely through their web browsers. We report on the substantial improvements in the speed and cost of data capture we have observed with this crowd-sourced paradigm. We also analyze a range of data quality factors by comparing a crowd-sourced data set involving 196 remote users to a smaller but more quality controlled lab-based data set. We focus our comparison on aspects that are especially important in our spoken dialogue research, including audio quality, the effect of communication latency on the interaction, our ability to synchronize the collected data, our ability to collect examples of excellent game play, and the naturalness of the resulting interactions. This analysis illustrates some of the current trade-offs between lab-based and crowd-sourced spoken dialogue data.

Keywords Crowd-sourcing • Web-based spoken • Dialogue system dialogue data collection

18.1 Introduction

In recent years, dialogue system researchers have been attracted to crowd-sourcing approaches for a number of data collection tasks that support system training and evaluation. Some of the tasks that have been explored include transcription (Parent and Eskenazi 2010), capture of speech and text for training language models (Liu et al. 2010), eliciting utterance texts that correspond to specific semantic forms (Wang et al. 2012), collecting text templates for generation (Mitchell et al. 2014), and collecting survey-style judgments about a dialogue system's performance (Yang et al. 2010). Crowd-sourcing and online data capture approaches have also been used

---

R. Manuvinakurike (✉) • D. DeVault
USC Institute for Creative Technologies, Playa Vista, Los Angeles, CA 90094, USA
e-mail: manuvinakurike@ict.usc.edu; devault@ict.usc.edu

G. Geunbae Lee et al. (eds.), *Natural Language Dialog Systems and Intelligent Assistants*, DOI 10.1007/978-3-319-19291-8_18

to collect interactive dialogues in which a single user interacts with a live dialogue system (e.g., Meena et al. 2014; Liu et al. 2010; Jiang et al. 2014).

We present in this paper a web framework that supports crowd-sourced collection of spoken dialogue interactions between two remote participants. To the best of our knowledge, this is the first time that crowd-sourcing has been applied to the collection of spoken dialogue interactions between two remote participants in support of dialogue system research. Crowd-sourcing has been used to collect text-based chat dialogues between remote participants; see, e.g., Lasecki et al. (2013). Such human–human dialogue data can be quite valuable, especially in the early stages of designing and building a dialogue system. Human–human data provides examples of domain-specific language and interaction that can inform a range of architecture and design choices in system building, as well as serving as initial training data for system components (Lasecki et al. 2013). The decision to collect spoken dialogues between human interlocutors online, rather than in a controlled lab setting, is a multifactorial one. Some of the important considerations include the introduction of browser-mediated interaction, limitations in available modalities for interaction, potential changes in demographics, data quality considerations, and the introduction of communication latency.

The research that motivates our crowd-sourced data collection involves fast-paced spoken dialogue games, in which interlocutors describe images to each other. An example interaction, drawn from the lab-based rapid dialogue game (RDG) corpus we previously collected (Paetzel et al. 2014), is shown in Fig. 18.1. In this excerpt, one player (the Director) tries to describe the image depicted with a red border at the top left of the screen to the other player (the Matcher), who sees the same array of eight images on his own screen but with their locations shuffled. The players are under substantial time pressure to complete as many images as they can within a fixed time limit. Natural spoken dialogue in this domain includes frequent overlapping speech (shaded in red), low latency turn-taking (as when the matcher asks *how many hands out?* and receives the answer *both hands* 215 ms later), mid-utterance repairs, interruptions, acknowledgments, and other low-latency responses. Capturing such rapid spoken exchanges over the internet presents a unique challenge. Particularly important factors for our dialogue system research, which aims to replicate these rapid dialogue skills in dialogue systems, include the quality of captured audio, the effect of communication latency on the interaction, the ability to collect examples of excellent game play, and naturalness of the interaction and turn-taking. In addition to describing our web framework, this paper presents a case study of how these factors differ between the lab-based corpus we previously collected and the crowd-sourced corpus we have collected with the new web framework.

18.2 The RDG-Image Game and Lab-Based Corpus

In the RDG-Image game (Paetzel et al. 2014), one person acts as a director (or "giver") and the other as a matcher (or "receiver"). Players are presented a set of eight images on separate screens. The set of images is exactly the same for both

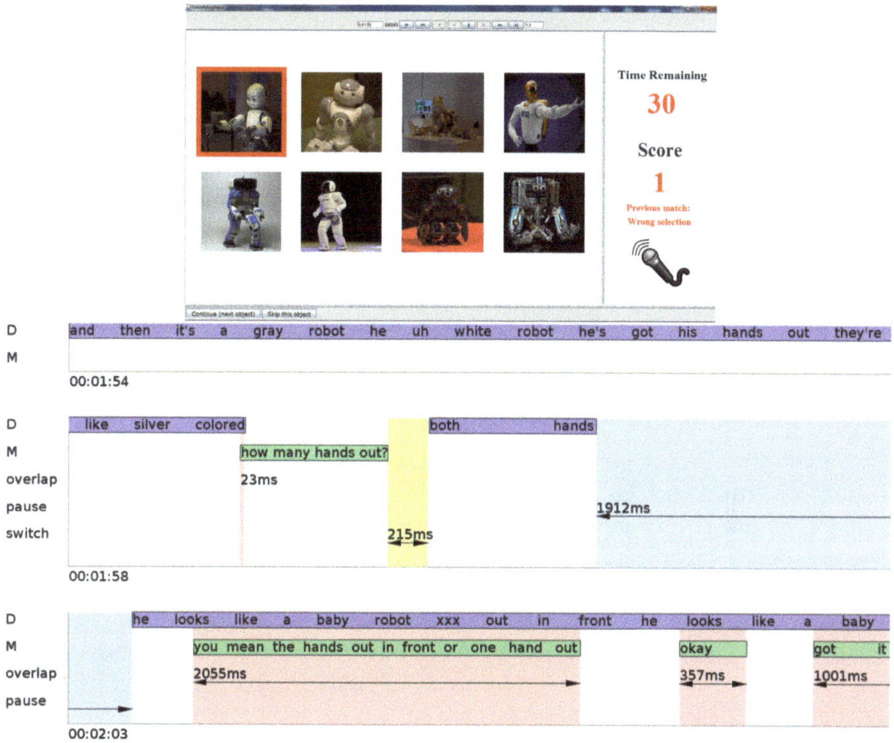

Fig. 18.1 An excerpt of human–human gameplay from our lab corpus. Segments of participant speech are arranged to show their temporal extents, with time increasing from left-to-right and from top-to-bottom. Speech is segmented at all silent pauses exceeding 300 ms. Periods of overlapping speech are shaded in *red*. Periods containing silent pauses by a single continuing speaker are shaded in *blue*. Periods of silence at speaker switches are shaded in *yellow*

players, but they are arranged in a different order on the screen. One of the images is randomly selected as a target image (TI) and it is highlighted on the giver's screen with a thick red border as shown in Fig. 18.1. The goal of the giver is to describe the TI so that the receiver is able to uniquely identify it from the distractors. Different categories are used for the image sets including pets, fruits, people wearing make-up, robots (Fig. 18.1), and castles, among others. When the receiver believes he has correctly identified the TI, he clicks on the image and communicates this to the giver who has to press a button to continue with the next TI. The team scores a point for each correct guess, with a goal to complete as many images as possible within each 140 s round. Each team participates in four main game rounds, with roles alternating between rounds.

Our lab-based corpus includes 64 participants (32 pairs) recruited on Craigslist. Our lab-based data collection protocol was carefully designed to capture multimodal data at high fidelity. The gestures and movements of each participant were recorded individually with Microsoft Kinect cameras and multiple Logitech webcams. (Note that the giver was told to provide clues only verbally, and the role of gesture is small in this game.) Audio was also recorded for each subject individually using high-quality Sennheiser microphones and Focusrite USB audio interfaces. Each participant's user interface was run on a separate lab computer. As the two computers were under our direct control, we were able to synchronize their system clocks using the Network Time Protocol (Mills et al. 2010). The two computers communicated over a gigabit ethernet connection, and all game events were logged with millisecond precision timestamps. This lab setup allowed us to synchronize all the collected game data with no observable synchronization challenges.

18.3 The Web-Based RDG-Image Game

To explore crowd-sourced data collection for RDG-Image, we adapted the Java-based RDG-Image user interface to a browser-based interface, shown in Fig. 18.2. The interface is broadly similar to the lab interface (Fig. 18.1). For the web-based game, we elected to pay a bonus to each player based on the number of correct images they achieve together within the time limit. To emphasize the monetary incentive, we display their score in terms of this bonus (marked "WINNINGS" in Fig. 18.2). The score is thus denominated in US Dollars rather than in points.

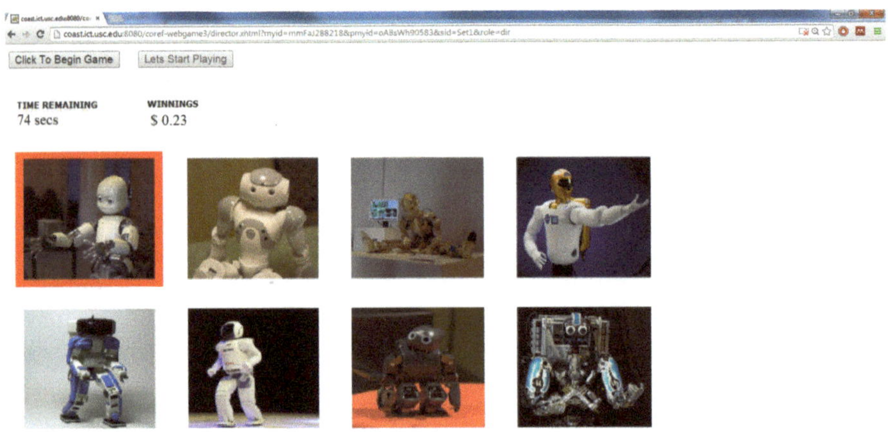

Fig. 18.2 The browser-based RDG-Image interface. This screenshot shows the director's browser, with the target image highlighted. The images in the matcher's browser appear in random order, and the matcher doesn't have the next question button. Otherwise the interface is the same

18.3.1 The Pair Me Up Web Framework

Pair Me Up is a software framework that supports web-based collection of spoken human–human dialogues between remote participants. It pairs consecutive web users together and connects them into a shared game session where they can converse freely and interact through their browsers. Pair Me Up leverages recent developments in web technologies that support development of web-based dialogue systems. It shares this approach with recent dialogue system research such as Jiang et al. (2014), which makes use of emerging web technologies to enable a spoken interaction between an individual remote web user and an automated dialogue system. In Pair Me Up, we use several of these new web technologies to build an interactive game where the servers can initiate events on remote client browsers, audio is streamed between two remote client browsers, and audio is captured to a server database.

The architecture of Pair Me Up is shown in Fig. 18.3. Two core technologies the system makes use of are websockets and webRTC (Vogt et al. 2013). Websockets enable two-way communication between the client and server, and they specifically enable the server to push events such as image set changes to the clients and the clients to send audio and game events such as button clicks to the server, without loading a separate URL. The streaming audio communication between the remote clients is currently set up using a separate SimpleWebRTC (http://simplewebrtc.com/) channel. The video channel is disabled for the current study due to bandwidth limitations observed in pilot testing and the fact that RDG-Image players primarily look at the images being described rather than each other.

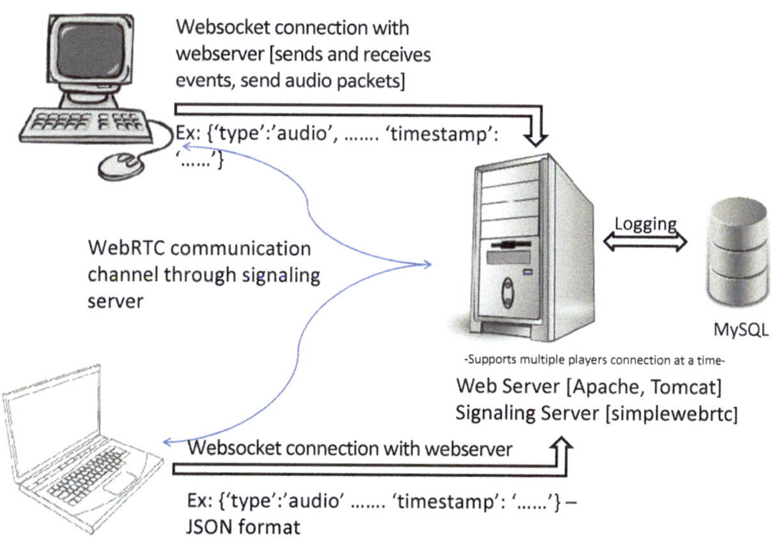

Fig. 18.3 Architecture of the pair me up system

18.3.2 Latency Measurement Protocol and Data Synchronization

In a lab-based study, network latency between machines can be minimized through use of high-speed LAN connections, and computer clocks can be synchronized using a method such as the Network Time Protocol (Mills et al. 2010). In a crowd-sourced data collection, network latency may be both higher and also harder to control. Additionally, security considerations rule out adjusting a remote user's system clock.

In our web-based game interface, latency can potentially affect the data we collect in several ways. There can be latency between when a remote user initiates an action in their UI and when the server learns that the action occurred, for example. Conversely, if the server initiates an event in the user's UI (e.g., changing the image set), this event may not actually occur in the user's UI until sometime later. Given the sensitivity of our research to having accurate timing of dialogue events, we implemented a simple latency measurement and synchronization protocol that allows us to (1) estimate the network latency between each client and the server, and (2) translate between timestamps collected from client machine system clocks and timestamps on our server.

Like the Network Time Protocol, our approach relies on the transmission of a series of request/response packets between the server and client machines. The protocol is illustrated in Fig. 18.4. At the beginning of each image set in the game, a request packet is sent from the server S to the remote client C. We denote the server's timestamp when this request is sent by t_S^a, using a subscript for the machine (S or C) whose clock generates the timestamp and a superscript (a, b, c, or d) for the four sequential events that occur as the request and response are sent and received by each machine. As part of the exchange, Pair Me Up code running in client C's browser computes a client system timestamp t_C^c and immediately sends this value with its response back to the server. The server receives the response at t_S^d. With each request/response cycle, the server therefore has a measure of the round trip latency

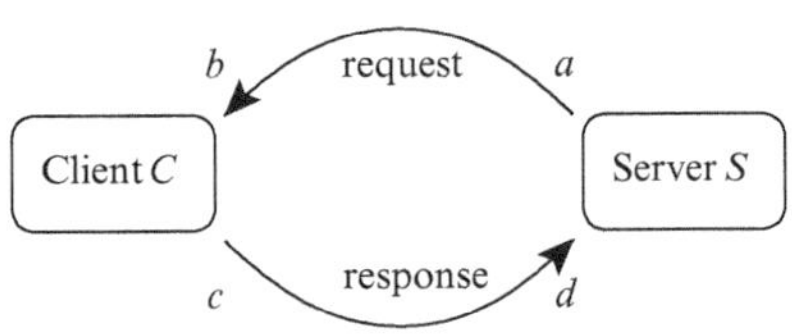

Event	Known time of event on S	Known time of event on C
a : Server S sends request	t_S^a	
b : Client C receives request		t_C^b
c : Client C sends response		t_C^c
d : Server S receives response	t_S^d	

Fig. 18.4 Latency measurement protocol

of server–client communication: roundtrip $= t_S^d - t_S^a$. Over the course of a game this request/response cycle happens in the background many times between the server and each of the remote clients. In order to relate client event timestamps to server event timestamps, we adopt the assumption that the client initiated its response at the midpoint of the server's observed roundtrip time:

$$t_S^c = \frac{1}{2}(t_S^a + t_S^d)$$

This provides us with a series of timestamp pairs, t_C^c and t_S^c, for the same event expressed on the client and system clocks. We then use a linear regression to estimate a translation between any arbitrary client timestamp t_C^e for event e and the corresponding server timestamp t_S^e:

$$t_S^e = w_1 \cdot t_C^e + w_2 \tag{18.1}$$

Of course, this translation can also be used bidirectionally to translate server timestamps into corresponding client timestamps. To enable approximate synchronization of all the collected data, all events originating on the two remote clients (including user mouse clicks and image selections, button presses, page request times, connection to partner, and audio chunk capture) are logged into the database with the associated client timestamps. Events originating on the server (including image set changes, countdown timer events, and score changes) are logged into the database with the associated server timestamps. All data and events are later approximately synchronized by translating all events onto a common timeline using Eq. (18.1). We can reconstruct all the events on the server timeline or user timeline as desired. One limitation of our current approach is that network latency is not completely constant, and thus a dynamic translation might achieve a more accurate synchronization.

18.4 Crowd-Sourced Data Set

We recruited 196 individuals from Amazon Mechanical Turk (AMT) to participate in the web-based RDG-Image game. The requirements to participate in the HIT were: (1) a high speed internet connection (5 mbps download, 2 mbps upload); (2) the latest Google Chrome web browser; (3) task acceptance of $\geq 92\,\%$; (4) previous participation in at least 50 HITs; (5) physical location in the United States; (6) must be a native English speaker; (7) must have a microphone; and (8) must not be on a mobile device. As part of self-qualifying themselves, Turkers verified their internet connection speed using the speedtest.net web service. Additionally, although this was not a strict requirement, they were strongly encouraged to listen to their computer's sound output using headphones rather than speakers. This instruction was added after pilot testing, to help reduce audio quality issues related to echo.[1] After self-qualifying for the HIT, users proceeded to the instructions, which were provided in both text and video format. The instruction video explained the interface in detail. Users then followed a link and waited until they were paired up with another Turker as a game partner. Access to each Turker's microphone and speakers was then requested from the users. The users then made sure their audio connection worked well. Before playing the game, they were shown a leaderboard where they could see how prior teams performed. After the game, they returned to the AMT site for a post-game questionnaire.

During the data collection, pairing of participants happened on a "first come, first served" basis. Pair Me Up simply connected each player to the next player who reached the same stage in the HIT, without any scheduling. To attract users, we posted a number of HITs and waited until two consecutive Turkers could be paired up to play the game. Our Pair Me Up server is currently able to support at least 12 simultaneous players (6 simultaneous games). We observed that this approach worked well provided that a sufficient number of HITs were made available on Mechanical Turk. However, we avoided posting too many HITs at once to prevent exceeding our server's capacity. When too few HITs were available, waiting times for a partner increased.

18.5 Results

18.5.1 Data Collection Throughput

In total our web-based data collection took place over 17 days and included 177 h of aggregate HIT time by 196 Turkers. We expect each HIT to take a minimum of about 15 min to complete, including reading instructions, 9 min and 20 s of actual gameplay, and the post-game questionnaire. The median time was nearly 38 min, which is about the same amount of time it took for the participants in the lab to complete the RDG-Image game and fill out all questionnaires. Most of the time spent by our web study participants was spent waiting for a partner. In future work we would like to reduce this wait time by pairing up partners more efficiently. The main bottleneck to parallel game collection on our server is the actual live gameplay, which requires transmission and logging of speech streams. Because our server can support at least six simultaneous live games, and the actual dialogue gameplay requires only 9 min and 20 s per pair, this translates into a potential data collection throughput for the Pair Me Up framework on a single server of hundreds of spoken dialogue games per day. In comparison, our lab-based data collection,

[1]When the users listen through speakers, it often happens that one of their microphones picks up the speech output of their partner, and echo ensues. We currently do not attempt to cancel this echo.

which yielded 32 subject pairs, took about a month to orchestrate and complete, due largely to the overhead of individual subject recruitment and scheduling, as well as the impossibility of parallelism given lab resources.

18.5.2 Audio Quality

In our lab-based corpus, audio was captured using high-quality microphones and audio hardware, which were calibrated and adjusted by lab staff for each participant. Additionally, our lab is generally a low noise environment that is free of most distractions. By contrast, in our web audio data, we have very little control over the participants' audio hardware and ambient environments. We observed captured audio to include a wide range of background noises, including televisions, cats meowing, dogs barking, and mobile phones ringing, among other distractions. Our primary use for this audio is through transcription and automatic speech recognition (ASR), in support of dialogue system research and development. We therefore assess audio quality by way of its suitability for these purposes. We currently have transcribed a subset of the web-based speech amounting to several hundred utterances. For this subset, despite the variable audio conditions, we have encountered no difficulties in transcribing the speech. To assess the audio quality in relation to ASR, we selected a random sample of 105 utterances each from the web corpus and the lab corpus. As part of transcription, these utterances were segmented from surrounding speech (using silence regions exceeding 300 ms) and manually transcribed. We then evaluated ASR word error rate for both samples using Google's ASR (https://www.google.com/speech-api/v2/recognize), a broad-coverage cloud-based industry speech recognizer which we have observed to have competitive performance in recent ASR evaluations for dialogue systems at our institute (Morbini et al. 2013). In our corpora, the observed word error rate (WER) of 24.10 in ASR for web-based audio is significantly higher ($W = 4647.5$, p-value $= 0.04285$, Wilcoxon rank sum test) than the WER of 19.83 for lab-based audio. This increase in WER of 4.27 for web-based audio provides perspective on the trade-offs between controlled lab-based audio capture and crowd-sourced online audio capture for dialogue system researchers.

18.5.3 Effect of Latency on Game Performance and Synchronization

We summarize the network latency for each user using the round trip time observed in the latency measurement protocol described in Sect. 18.3.2. Higher values indicate higher network latency that could adversely impact various aspects of gameplay, for example UI responsiveness to user button clicks as well as the speech channel. We observed a mean roundtrip latency of 136.9 ms (median 108.0 ms, standard deviation 84.9 ms, min 29.0 ms, max 464.0 ms). To understand how latency affects overall game performance, we investigated the relationship between roundtrip latency and score (number of correct images). We observed a slight weak, but significant, negative correlation between latency and score ($r = -0.16, p < 0.05$). Upon closer examination, the negative effect of latency on score seems to be limited to those players with relatively high latency. We find no significant correlation between score and latency for players whose latency is below 250 ms ($r = -0.06, p = 0.44$). Comparing the population of low latency players (latency $<=$ 250 ms, $N = 177$, mean score 50.7) to high latency players (latency $>$ 250 ms, $N = 19$, mean score 40.5), we observe a significant difference in scores ($p < 0.05$, Wilcoxon rank-sum test). We interpret these data as suggesting that if latency is low enough, its effect on game score is negligible. Additionally, we used our latency measurement and synchronization protocol to construct more than 20 synchronized videos that combine the two users' speech streams with a recreation of each user's UI state at each moment (including images observed, button clicks, etc.). If timeline correction using Eq. (18.1) is not performed, such videos exhibit numerous clear synchronization problems. After timeline correction, we have found that the combined videos appear remarkably well synchronized. Upon observation of these videos, we are unable to detect any remaining latency or synchronization issues, and we view the data as sufficiently synchronized for our research purposes. We would like to further investigate the exact accuracy of data synchronization achieved in future work.

18.5.4 Cost, Gameplay Quality, and Naturalness of Interaction

We summarize the study cost, scores attained, and basic demographic data for our two corpora in Table 18.1. From Table 18.1 we can see that the web-study data is 7.8× less expensive per participant to collect (once the Pair Me Up infrastructure is in place). In terms of acquiring examples of excellent gameplay, which is one of our research requirements, we found that our web-study players scored significantly higher than the players in lab ($W = 5389$, $p = 0.01875$, Wilcoxon rank sum test). The full explanation for this difference is unclear as there were several differences between the web study and the lab study. One difference is

Table 18.1 Cost, scores attained, and demographic data for our web and lab studies

	Web	Lab
N	196	64
Average pay per player	$1.915	$15
Scores (%) [mean, SD, min, max, median]	49.8, 13.1, 22, 78, 51	45, 13.0, 20, 68, 44
Age (%) [mean, SD, min, max, median]	31.3, 8.2, 19, 68, 29	36.6, 12.7, 18, 60, 34.5
Gender (%) [female, male]	53.3, 46.7	55, 45

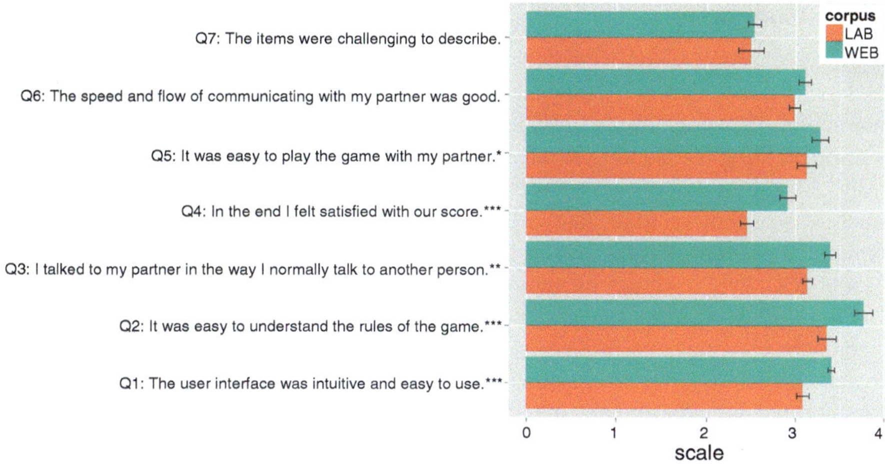

Fig. 18.5 Subjective questionnaire results for questions related to interaction naturalness and usability of user interface. Means and standard errors are shown for all questions (*$p < 0.05$, **$p < 0.014$, ***$p < 0.001$)

that web-study participants were incentivized with a bonus payment per correct image, while lab study participants were paid a flat rate of $15 for participation. Demographic differences between Turkers and Los Angeles area Craigslist users may also have played a role; for example, our web-study participants were younger on average. In any case, we conclude that it is possible to collect examples of excellent gameplay for RDG-Image with a crowd-sourced data collection. All participants filled out post-game subjective questionnaires, providing answers on a 5-point Likert scale. We were especially interested in the perceived naturalness of the interaction and the usability of the interface, and we present several observations in Fig. 18.5. All significance tests are Wilcoxon rank sum tests.

Web-study participants gave significantly higher ratings of the user interface being intuitive and easy to use (Q1). They also gave higher ratings to the ease of understanding the game rules (Q2) and it being easy to play the game with their partner (Q5). These findings may be partially explained by the more detailed instructions we provided for web users about the browser interface, including the addition of video-based instructions. Demographic differences and possible comfort in using a browser-based interface could potentially play a role as well. In terms of naturalness of the interaction, the results were also favorable for the web-based study. Despite our concern about network latency affecting interaction naturalness, we observed no significant difference in ratings of the speed and flow of communication between the web study and the lab study (Q6). In fact, web-study participants gave significantly higher ratings to it being easy to play the game with their partner (Q5), satisfaction with their score (Q4), and a rating of whether they spoke the way they normally do with the partner they were paired with (Q3). The fact that web-study participants scored higher than lab-study participants may play a role in the perceived ease of playing with their partner and score satisfaction.

18.6 Conclusions

We have presented a web framework called Pair Me Up that enables spoken dialogue interactions between remote users to be collected through crowd-sourcing. We have confirmed, for spoken dialogue interactions in the RDG-Image game, the commonly observed pattern that crowd-sourced data collection over the web can be faster and much less expensive than data collection in the lab. At the same time, we have explored several trade-offs in web-based vs. lab-based data collection for dialogue system research. In terms of audio quality, we have found an increase of about 4 % in ASR word error rate for web-based audio data. Such an increase may be acceptable by many researchers in exchange for easier data collection. In terms of network latency, we have found that while it is an important consideration, it does not rule out natural real-time dialogue between the remote participants and that data can still be synchronized sufficiently for our purposes using a straightforward latency measurement protocol. We have observed that the quality of gameplay, as determined by scores achieved and several subjective assessments by Turkers, was higher for our crowd-sourced study than in the lab.

Acknowledgements We thank Maike Paetzel. This material is based upon work supported by the National Science Foundation under Grant No. IIS-1219253. Any opinions, findings, and conclusions or recommendations expressed in this material are those of the author(s) and do not necessarily reflect the views of the National Science Foundation. For the images in Figs. 18.1 and 18.2, we thank (numbering 1–8 from left-right, top-bottom): [1,2] Jiuguang Wang (CC BY-SA 2.0), [3] Alex Haeling (CC BY 2.0), [4] NASA, [5] Joe Wu (CC BY-NC-SA 2.0), [6] RoboCup2013 (CC BY-NC-SA 2.0), [7] Waag Society (CC BY 2.0), and [8] Janne Moren (CC BY-NC-SA 2.0).[2]

References

Jiang R, Banchs RE, Kim S, Yeo KH, Niswar A, Li H (2014) Web-based multimodal multi-domain spoken dialogue system. In: Proceedings of 5th international workshop on spoken dialog systems

Lasecki W, Kamar E, Bohus D (2013) Conversations in the crowd: collecting data for task-oriented dialog learning. In: Human computation workshop on scaling speech and language understanding and dialog through crowdsourcing

Liu S, Seneff S, Glass J (2010) A collective data generation method for speech language models. In: Spoken language technologies workshop (SLT)

Meena R, Boye J, Skantze G, Gustafson J (2014) Crowdsourcing street-level geographic information using a spoken dialogue system. In: The 15th annual SIGdial meeting on discourse and dialogue (SIGDIAL)

[2] [1] http://www.flickr.com/photos/jiuguangw/4981810943/, [2] http://www.flickr.com/photos/jiuguangw/4982411246/, [3] http://www.flickr.com/photos/alexhealing/2841176750/, [5] http://www.flickr.com/photos/ozzywu1974/325574892/, [6] http://www.flickr.com/photos/robocup2013/9154156312/, [7] http://www.flickr.com/photos/waagsociety/8463802099/, and [8] http://www.flickr.com/photos/jannem/1885853738/.

Mills D, Martin J, Burbank J, Kasch W (2010) Network time protocol version 4: protocol and algorithms specification. http://www.ietf.org/rfc/rfc5905.txt
Mitchell M, Bohus D, Kamar E (2014) Crowdsourcing language generation templates for dialogue systems. In: the 8th international natural language generation conference (INLG)
Morbini F, Audhkhasi K, Sagae K, Artstein R, Can D, Georgiou P, Narayanan S, Leuski A, Traum D (2013) Which ASR should I choose for my dialogue system? In: Proceedings of the SIGDIAL 2013 conference, Metz
Paetzel M, Racca DN, DeVault D (2014) A multimodal corpus of rapid dialogue games. In: Language resources and evaluation conference (LREC)
Parent G, Eskenazi M (2010) Toward better crowdsourced transcription: transcription of a year of the let's go bus information system data. In: IEEE workshop on spoken language technology
Vogt C, Werner MJ, Schmidt TC (2013) Leveraging webrtc for p2p content distribution in web browsers. In: International Conference on Network Protocols (ICNP), pp 1–2
Wang W, Bohus D, Kamar E, Horvitz E (2012) Crowdsourcing the acquisition of natural language corpora: methods and observations. In: Spoken language technology workshop (SLT), pp 73–78
Yang Z, Li B, Zhu Y, King I, Levow G, Meng H (2010) Collection of user judgments on spoken dialog system with crowdsourcing. In: Spoken language technologies workshop (SLT)

Chapter 19
Creating a Virtual Neighbor

Carina Corbin, Fabrizio Morbini, and David Traum

Abstract We present the first version of our Virtual Neighbor, who can talk with users about people employed in the same institution. The Virtual Neighbor can discuss information about employees in a medium-sized company or institute with users. The system acquires information from three sources: a personnel directory database, public web-pages, and through dialogue interaction. Users can interact through face-to-face spoken dialogue, using components from the ICT Virtual human toolkit, or via a chat interface.

Keywords Natural language generation • Virtual directory • Natural language understanding • Dialogue management • System-user interaction

19.1 Introduction

The Virtual Neighbor project strives to recreate the conversations neighbors often share. When we want to learn more about those who live around us, most people have a neighbor they can turn to for an update on the latest information. In this paper, we present first results in the creation of a Virtual Human that could do the same thing. A virtual neighbor shares much with a personal assistant dialogue system, such as Siri or Cortana, in terms of being able to answer a range of information questions, however the virtual human has a more persistent presence (beyond just answering an isolated query) and is aimed at finding out more about you, as well as just answering your questions.

Miki, the Virtual Human Neighbor, was designed to deal initially with topics related to co-worker information. She can answer questions that range from where someone is located to the last project he or she worked on. Miki can answer questions, prompt interlocutors with suggestions, and learn information from interlocutors. Miki does this with the help of three sources of information.

C. Corbin (✉) • F. Morbini • D. Traum
Institute for Creative Technologies, University of Southern California,
Los Angeles, CA 90094, USA
e-mail: ccorbin17@amherst.edu; morbini@ict.usc.edu; traum@ict.usc.edu

G. Geunbae Lee et al. (eds.), *Natural Language Dialog Systems and Intelligent Assistants*, DOI 10.1007/978-3-319-19291-8_19

She has a main knowledge base that contains the basic directory information about every employee, in the form of a spreadsheet. She connects to a web crawler that searches through a company or institute's website and gathers information such as who worked on which projects. Lastly, she learns from the user.

19.2 Related Work

There are a number of dialogue systems that provide directory assistance information, for example Yoshioka et al. (1994), Seide and Kellner (1997), Buntschuh et al. (1998), de Córdoba et al. (2001). These mostly focus on the challenges for ASR in dealing with a large database of names and handle only contact information for the individuals. Our problem is focused a bit differently, in a medium-sized work environment (e.g., 100–200 people), where there is limited ambiguity between names, but where people also want to talk about other kinds of information related to the people at work, including who their supervisor is and what they work on.

Our system is integrated within the ICT virtual human toolkit (Hartholt et al. 2013). This provides connections to various speech recognition and synthesis options, as well as a visual environment and the virtual human body and nonverbal behavior.

For natural language understanding, generation, and dialogue management, we built on top of the FLoRes system (Morbini et al. 2012). This system allowed us to define the information state for employees and attributes and a set of policies for prompting and responding to the user. In addition, related NLU and NLG routines allowed us to specify a set of speech acts, parameters, and language templates for this domain.

19.3 System Overview

Figure 19.1 shows a screenshot of the Miki character. Users can talk to her face to face, using any toolkit-compatible speech recognizer (Morbini et al. 2013) (we tested with apple dictation and pocketsphinx). Additionally, it is possible to text miki using a chat interface.

The system's knowledge is kept in the form of a database, with several types of information (people, locations, titles, projects) and several relations: for each individual, their first and last names, job title, their office location, and which projects they've worked on. Most of the information comes from an employee directory database. This database is used to power an internal web directory, but can also be saved as an excel spreadsheet. The virtual human neighbor uses the spreadsheet, since it can also easily be added to by hand.

There are two other Excel sheets used to allow the Virtual Human Neighbor to respond realistically. The first is the System Utterances sheet. This sheet gives the system different options for realizing speech acts with no parameters, such as

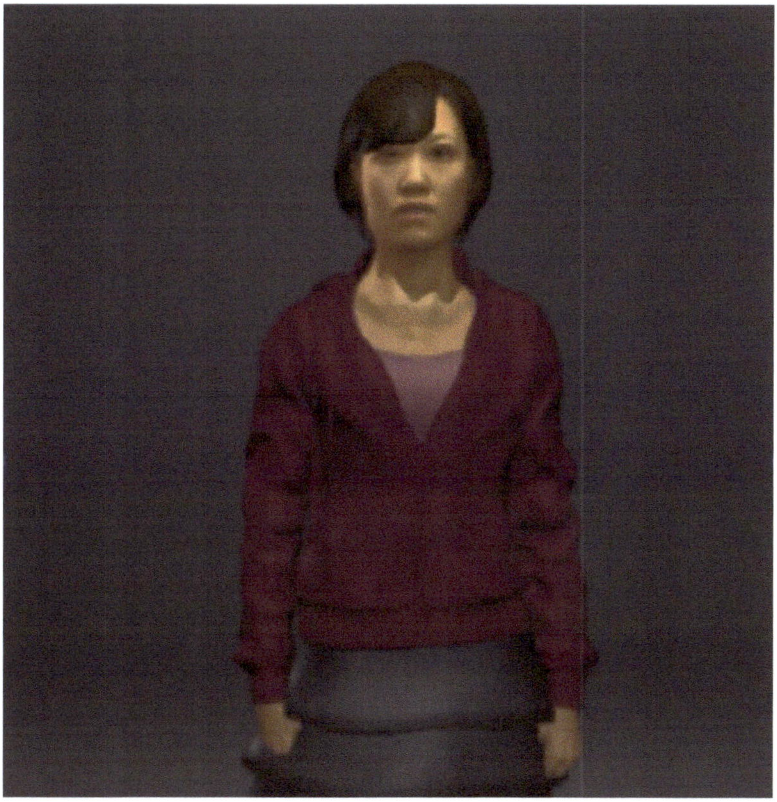

Fig. 19.1 Virtual human neighbor

Fig. 19.2 A typical first interaction with the virtual neighbor

2.1	User:	Hello?
2.2	Miki:	Hi, I'm Miki, how may I help you?
2.3	User:	What can I ask you?
2.4	Miki:	You can ask me about your supervisor.

greetings and commonly asked questions about the system itself. The sheet gives the system several different ways to say hello, goodbye, and several suggestions to help the user interact with the system. Our other Excel sheet lists potential user utterances and serves as training data for our natural language understanding module. The goal of the NLU module is to recognize the user's speech act and parameters. Training data can be given, including named entities (such as first name or office location), and the NLU abstracts to other entities of the same type. Figure 19.2 shows an example of the System Utterances in real conversation.

System utterances with parameters that can be filled in from the information state use a natural language generator rather than the system utterances spreadsheet. For example, in Fig. 19.3, utterance 3.1 is a query about a person with first name parameter filled in. The system has multiple people with that first name and generates a list output in utterance 3.2. When the last name is added in utterance

3.1 User: Who is David?
3.2 Miki: We have many people who share that name: David Traum, David Pynadeth, David DeVault,...
3.3 User: Who is David Traum?
3.4 Miki: David Traum works at ICT as a Director of the Natural Language Dialogue Group
3.5 User: Where can I find him?
3.6 Miki: You can visit David at Rm 350

Fig. 19.3 Dialogue with named entity recognition and generation

4.1 User: Where does Carina sit?
4.2 Miki: You can visit Carina off site
4.3 User: I am Carina
4.4 Miki: I know you, Carina. You are a Summer Intern-R E U
4.5 User: I now sit at 357.0
. . .
4.6 User: Where is Carina's office?
4.7 Miki: You can visit Carina at 357

Fig. 19.4 Acquiring information from users

5.1 User: What projects did Morbini work on?
5.2 Miki: Fabrizio worked on 2 prototypes including Combat Hunter Action and Observation Simulation (CHAOS).

Fig. 19.5 Acquiring information from websites

3.3, a unique record is found, and the title is returned in utterance 3.4 (using the first and last name parameters). Utterance 3.5 illustrates an ability to do anaphora resolution to the person currently in focus. Utterance 3.6 uses only the first name parameter as well as the requested information about this person.

In a true conversation, humans don't just ask questions and receive answers. We alternate between being the inquirer and the informant, something we want the Virtual Neighbor to do as well. The storing of information about the subject also allows the system to be updated through user interaction. Let's say the knowledge base does not have the office location of an employee. A person can approach the system and state where the person sits. The person in question can also introduce herself and then state where she sits using this same process. This makes interacting with the Virtual Neighbor a more well-rounded experience. Figure 19.4 illustrates this process, updating a location, which can be retrieved later.

To extend the topics of conversation, we also created a webcrawler that scrapes the organization's public webpages for more information about each employee. Our first example is acquiring information about which prototypes were worked on, using a page that lists institute prototypes (http://ict.usc.edu/prototypes/all/). Each of the linked pages has a "team" section, with a list of employees who worked on the prototype. This enables dialogue like that in Fig. 19.5.

19.4 Experimental Design

To evaluate the system, our main goal is to analyze how comfortable people feel interacting with the Virtual Neighbor: to what extent people actually could see themselves using the Virtual Neighbor as a common resource. We plan to test this, by having participants fill out a Pre-Test to see if the user commonly uses the employee directory, visits the company webpages and has an interest in working with the Virtual Neighbor. Next, we will give each participant a set of tasks, such as looking up information in the employee directory. Then we will ask participants to get the same information from the Virtual Neighbor by asking the system given questions. We will leave a few questions open ended, for example Participants will be invited to ask the location or title of another employee in the office of their choice. After the experiment, we will give each participant a Post-Test. This asks the participant to rank their experience with both the employee directory and the Virtual Neighbor. We will also query their favorite and least favorite parts about interacting with the system.

19.5 Conclusion

In this paper, we present the first version of our Virtual Human Neighbor, who can talk with users about people employed in the same institution. The Virtual Human neighbor can acquire information from three sources: a personnel directory database, public webpages, and through dialogue interaction. Future work involves completing the evaluation described in the previous section, as well as adding more kinds of information. We would also like to include other kinds of speech acts and dialogue games about this material, such as quizzes to get to know more about other employees, or gossip (Brusk et al. 2010).

Acknowledgements The first author was supported by the National Science Foundation under grant 1263386, "REU Site: Research in Interactive Virtual Experiences" (PI: Evan Suma).

The effort described here has been sponsored by the U.S. Army. Any opinions, content or information presented do not necessarily reflect the position or the policy of the United States Government, and no official endorsement should be inferred.

We would like to thank Ed Fast and Eli Pincus for help with this work.

References

Brusk J, Artstein R, Traum D (2010) Don't tell anyone! two experiments on gossip conversations. In: 11th SIGdial workshop on discourse and dialogue, Tokyo, pp 193–200

Buntschuh B, Kamm C, Fabbrizio GD, Abella A, Mohri M, Narayanan SS, Zeljkovic I, Sharp RD, Wright JH, Marcus S, Shaffer J, Duncan R, Wilpon JG (1998) Vpq: a spoken language interface to large scale directory information. In: Proceedings of InterSpeech, Sydney, pp 2863–2867 de Córdoba R, San Segundo R, Montero JM, Colás J, Ferreiros J, Guarasa JM, Pardo JM (2001) An interactive directory assistance service for Spanish with large-vocabulary recognition. In: INTERSPEECH, pp 1279–1282

Hartholt A, Traum D, Marsella SC, Shapiro A, Stratou G, Leuski A, Morency LP, Gratch J (2013) All together now: introducing the virtual human toolkit. In: International conference on intelligent virtual humans, Edinburgh

Morbini F, DeVault D, Sagae K, Gerten J, Nazarian A, Traum D (2012) Flores: a forward looking, reward seeking, dialogue manager. In: Proceedings of international workshop series on spoken dialogue systems technologies (IWSDS-2012)

Morbini F, Audhkhasi K, Sagae K, Artstein R, Can D, Georgiou P, Narayanan S, Leuski A, Traum D (2013) Which ASR should i choose for my dialogue system? In: Proceedings of the SIGDIAL 2013 conference. Association for Computational Linguistics, Metz, pp 394–403. http://www.aclweb.org/anthology/W/W13/W13-4064

Seide F, Kellner A (1997) Towards an automated directory information system. In: Proceedings of Eurospeech, pp 1327–1330

Yoshioka O, Minami Y, Shikano K (1994) A multi-modal dialogue system for telephone directory assistance. In: International conference on spoken language processing (ICSLP'94)

Chapter 20
Decision Making Strategies for Finite-State Bi-automaton in Dialog Management

Fabrizio Ghigi and M. Inés Torres

Abstract Stochastic regular *bi-languages* has been recently proposed to model the joint probability distributions appearing in some statistical approaches of spoken dialog systems. To this end a deterministic and probabilistic finite-state *bi-automaton* was defined to model the distribution probabilities for the dialog model. In this work we propose and evaluate decision strategies over the defined probabilistic finite-state *bi-automaton* to select the best system action at each step of the interaction. To this end the paper proposes some heuristic decision functions that consider both action probabilities learn from a corpus and number of known attributes at running time. We compare heuristics either based on a single next turn or based on entire paths over the automaton. Experimental evaluation was carried out to test the model and the strategies over the Let's Go Bus Information system. The results obtained show good system performances. They also show that local decisions can lead to better system performances than best path-based decisions due to the unpredictability of the user behaviors.

Keywords Statistical dialog management • Decision strategies

20.1 Introduction

Spoken dialog systems (SDS) enable human–machine interaction using natural spoken language (Raux et al. 2005; Seneff and Polifroni 2000). The process of interaction between the machine and a real user passes through several steps. One of the crucial steps in this process is the election of a next system action, a task performed by the dialog manager (DM). The DM is the module responsible of pursue the dialog goal by choosing a coherent action in response to a user input (Churcher et al. 1997). Due to its complexity the design of DM has been traditionally based on hand-crafted rules (Lee et al. 2006; Bohus and Rudnicky 2009). However, over the last few years, approaches that use statistical frameworks to deal with decision strategies

F. Ghigi (✉) • M.I. Torres
Departamento Electricidad y Electrónica, Universidad del País Vasco, Bilbao, Spain
e-mail: fabrizio.ghigi@ehu.es; manes.torres@ehu.es

G. Geunbae Lee et al. (eds.), *Natural Language Dialog Systems and Intelligent Assistants*, DOI 10.1007/978-3-319-19291-8_20

and task models have been providing compelling results on modeling interaction. These include Bayesian networks (Thomson et al. 2010), Stochastic Finite-State models (Hurtado et al. 2010; Inés Torres 2013), and the state-of-the-art Partially Observable Markov Decision Process (Williams and Young 2007; Jurčíček et al. 2012). The interactive pattern recognition framework (Toselli et al. 2011) has also been proposed to represent SDS (Inés Torres et al. 2012). This formulation needs to estimate the joint probability distribution over the semantic language provided by the speech understanding system and the language of actions provided by the DM. In a previous work (Inés Torres 2013) we have proposed to model this joint probability distribution by stochastic regular *bi-languages*. To this end a deterministic and probabilistic finite-state bi-automata (PFSBA) was defined in that work. Our goal now in this paper is to propose and evaluate DM strategies over this PFSBA-based dialog model. We are aimed at providing the DM with the best decision at each system turn. This decision will be selected according to some heuristic search on the model graph at running time. In Sect. 20.2 we summarize the deterministic PFSBA defined in Inés Torres (2013). In Sect. 20.3 we propose four decision strategies to be implemented by the DM at each system turn. The experiments and the results obtained are described in Sect. 20.4. Then Sect. 20.5 reports some final remarks and the future work planned.

20.2 Model Definition

Let us consider an SDS as an interactive pattern recognition system (Toselli et al. 2011; Inés Torres et al. 2012). Let now h be a hypothesis or output that the dialog manager of an SDS proposes. Then the user provides some feedback signals, f, which iteratively help the dialog manager to refine or to improve its hypothesis until it is finally accepted by the user. A basic simplification is to ignore the user feedback except for the last interaction a hypothesis h'. Assuming the classical *minimum-error criterion* Bayes' decision rule is simplified to maximize the posterior $Pr(h|h',f)$, and a best hypothesis $\hat{h}$ is obtained as follows:

$$\hat{h} = \arg\max_{h \in \mathscr{H}} P(h|h',f) \tag{20.1}$$

This maximization procedure defines the way the dialog manager of an SDS chooses the best hypothesis, i.e. the best action at each interaction step, given the previous hypothesis h' and the user feedback f. However, alternative criteria could also be considered as shown in Sect. 20.3. In an SDS, the interpretation of the user feedback f cannot be considered a deterministic process. In fact the space of decoded feedback $\mathscr{D}$ is the output of an ASR system. Thus a best hypothesis can be obtained as follows (Toselli et al. 2011; Ghigi et al. 2013; Inés Torres et al. 2012):

$$\hat{h} = \arg\max_{h \in \mathscr{H}} \sum_{d \in \mathscr{D}} P(h, d|h', f) \tag{20.2}$$

where f is the user turn, d is the decoding of the user turn, h is the hypothesis or output produced by the system, and h' is the *history of the dialog*.

The user feedback f depends on its previous feedback f' according to some unknown distribution $P(f|f', h)$, which represents the user response to the history of system hypotheses and user feedbacks. This distribution considers the user behavior and stands for a user model $\mathscr{M}_{\mathscr{U}}$. However, feedback f' produced by the user in the previous interaction is not corrupted by any noisy channel, such as an ASR system, before arriving to the user again. Thus, a deterministic decoding $d : \mathscr{F} \to \mathscr{D}$ maps each user turn signal into its corresponding unique decoding $d' = d(f')$ before arriving to the user. Consequently the *best* user feedback $\hat{f}$ is the one that maximizes the posterior $P_{\mathscr{M}_{\mathscr{U}}}(f|d', h)$

$$\hat{f} = \arg\max_{f \in \mathscr{F}} P(f|d', h) \approx \arg\max_{f \in \mathscr{F}} P_{\mathscr{M}_{\mathscr{U}}}(f|d', h) \tag{20.3}$$

where $\hat{f}$ is estimated using only the hypothesis produced by the system and the feedback produced by the user in the previous interaction step according to its *user model*. Figure 20.1 shows some user-manager interaction steps.

We are now summarizing the probabilistic dialog model defined in Inés Torres (2013) to deal with both the dialog manager hypothesis probability distribution $P(h|d, h')$ and the user feedback probability distribution $P(f|h, d')$. Let Σ be the finite alphabet of semantic symbols provided by some speech understanding system. Thus, $\tilde{d}_i = d_1 \ldots d_{|\tilde{d}_i|} \in \Sigma^{\leq m}$ represents the decoding of a user feedback f. Let now Δ be the finite alphabet of dialog acts that compose each of the hypotheses $\tilde{h}_i = h_1 \ldots h_{|\tilde{h}_i|} \in \Delta^{\leq n}$ provided by the dialog manager. Let $\mathbf{z}$ be a *bi-string* over

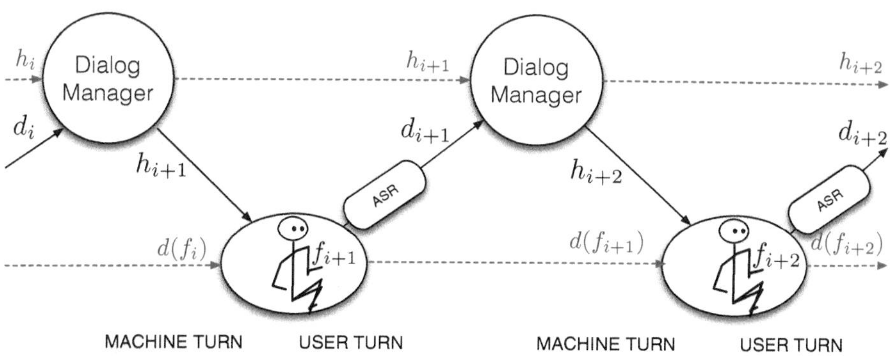

Fig. 20.1 User-manager interaction steps. h is the hypothesis produced by the system that depends on the previous hypothesis h' and the decoded user feedback d. f is the user turn that depends on h and on the previous user feedback f'

the extended alphabet $\Gamma \subseteq \Sigma^{\leq m} \times \Delta^{\leq n}$ such as $\mathbf{z} : \mathbf{z} = z_1 \ldots z_{|\mathbf{z}|}$, $z_i = (\tilde{d}_i : \tilde{h}_i)$ where $\tilde{d}_i = d_1 \ldots d_{|\tilde{d}_i|} \in \Sigma^{\leq m}$ and $\tilde{h}_i = h_1 \ldots h_{|\tilde{h}_i|} \in \Delta^{\leq n}$. A dialog model $\mathcal{DM}$ is defined as a deterministic and probabilistic finite-state *bi-automaton* $\mathcal{DM} = (\Sigma, \Delta, \Gamma, Q, \delta, q_0, P_f, P)$ where

- Σ and Δ are two finite alphabets representing semantic symbols provided by the user and dialog acts provided by the dialog manager, respectively, Γ is an extended alphabet such that $\Gamma \subseteq (\Sigma^{\leq m} \times \Delta^{\leq n})$, $m, n \geq 0$. ϵ represents the empty symbol for both alphabets, i.e., $\epsilon \in \Sigma$, $\epsilon \in \Delta$, and $(\tilde{\epsilon} : \tilde{\epsilon}) \in \Gamma$. To simplify let $\tilde{\epsilon}$ be ϵ.
- $Q = Q_{\mathcal{M}} \bigcup Q_{\mathcal{U}}$ is a finite set of states labelled by *bi-strings* $(\tilde{d}_i : \tilde{h}_i) \in \Gamma$. The set $Q_{\mathcal{M}}$ includes machine states before a machine turn providing a hypothesis and the set $Q_{\mathcal{U}}$ includes user states before providing a feedback.
- $\delta \subseteq Q \times \Gamma \times Q$ is the union of two sets of transitions $\delta = \delta_{\mathcal{M}} \bigcup \delta_{\mathcal{U}}$ as follows:
  - $\delta_{\mathcal{M}} \subseteq Q_{\mathcal{M}} \times \Gamma \times Q_{\mathcal{U}}$ is a set of transitions of the form $(q, (\epsilon : \tilde{h}_i), q')$ where $q \in Q_{\mathcal{M}}, q' \in Q_{\mathcal{U}}$ and $(\epsilon : \tilde{h}_i) \in \Gamma$
  - $\delta_{\mathcal{U}} \subseteq Q_{\mathcal{U}} \times \Gamma \times Q_{\mathcal{M}}$ is a set of transitions of the form $(q, (\tilde{d}_i : \epsilon), q')$ where $q \in Q_{\mathcal{U}}, q' \in Q_{\mathcal{M}}$ and $(\tilde{d}_i : \epsilon) \in \Gamma$
- $q_0 \in Q_{\mathcal{M}}$ is the unique initial state and it is labelled as $(\epsilon : \epsilon)$.
- $P_f : Q \to [0, 1]$ is the final-state probability distribution
- $P : \delta \to [0, 1]$ defines transition probability distributions $(P(q, b, q') \equiv Pr(q', b|q)$ for $b \in \Gamma$ and $q, q' \in Q)$ such that

$$P_f(q) + \sum_{b \in \Gamma, q' \in Q} P(q, b, q') = 1 \quad \forall q \in Q \tag{20.4}$$

  where a transition (q, b, q') is completely defined by q and b. Thus, $\forall q \in Q$, $\forall b \in \Gamma \ |\{q' : (q, b, q')\}| \leq 1$:

Let $\mathbf{z}$ be a *bi-string* over the extended alphabet $\Gamma \subseteq \Sigma^{\leq m} \times \Delta^{\leq n}$ such as $\mathbf{z} : \mathbf{z} = z_1 \ldots z_{|\mathbf{z}|}$, $z_i = (\tilde{d}_i : \tilde{h}_i)$. Let now $\theta = (q_0, z_1, q'_1, z_2, q_2, \ldots, q'_{|\mathbf{z}|-1}, z_{|\mathbf{z}|}, q_{|\mathbf{z}|})$, $q_i \in Q_{\mathcal{M}}$, $q'_i \in Q_{\mathcal{U}}$, be a path for $\mathbf{z}$ in $\mathcal{DM}$. The probability of generating θ is

$$Pr_{\mathcal{DM}}(\theta) = \left(\prod_{j=1}^{|\mathbf{z}|} P(q_{j-1}, z_j, q'_j) \right) \cdot P_f(q_{|\mathbf{z}|}) \tag{20.5}$$

$\mathcal{DM}$ is unambiguous. Then, a given *bi-string* $\mathbf{z}$ can only be generated by $\mathcal{DM}$ through a unique valid path $\theta(\mathbf{z})$. Thus, the probability of generating $\mathbf{z}$ with $\mathcal{DM}$ is $Pr_{\mathcal{DM}}(\mathbf{z}) = Pr_{\mathcal{DM}}(\theta(\mathbf{z}))$. Additionally, each machine and/or user state need to be labelled with the values of all relevant internal variables, which can be updated after each user turn. Thus, an additional alphabet appears to represent valued attributes of these internal variables, thus leading to an *attributed* model (Inés Torres 2013). These internal variables are a subset of the semantic decoding set, i.e., the subset of Σ set that consists of task-dependent symbols. These internal variables can lead to simple *known*, *unknown* attributes that can just be represented by the presence or absence of the attribute at each state. Thus, the new alphabet represents just the knowledge of the value. Alternatively confidence measures can also be considered. The model $\mathscr{DM}$ was then extended to add another finite alphabet Ω. Each state $q \in Q$ is now labelled by *bi-strings* $[(\tilde{d}_i : \tilde{h}_i), \tilde{w}_i] \in \Gamma \times \Omega$ where the valued attributes are also considered. The knowledge of the attributes leads to different strategies for the dialog manager since the transition function $\delta \subseteq Q \times \Gamma \times Q$ and the transition probability distribution $P : \delta \rightarrow [0, 1]$ have a strong dependency of internal attributed attached to the states.

Example 1. Let us take a dialog from Let's Go (Raux et al. 2005) task that was used in this work for experiments.

U:	I'm leaving from CMU
	PlaceInformation[DeparturePlace]
S:	Leaving from <query.departureplace CMU>. Is this correct?
	Explicit confirm
U:	Yes.
	Generic[Yes]
S:	Right. What is your destination?
	Inform : *confirmokay* *Request* : *queryarrival*

$\Sigma = \{PlaceInformation.DeparturePlace, Generic.Yes\}$ is the set of user symbols, $\Delta = \{Explicit.confirm, Inform.confirm\_okay, Request.query\_arrival\}$ is the alphabet of system dialog acts, and $\Omega = \{\text{query.departureplace}\}$ is the alphabet of the task attributes. Figure 20.2 shows a $\mathscr{DM}$ where bold lines define path θ matching some *bi-string*.

20.3 Heuristic Functions to Achieve User Goals

In this section we define four heuristic functions that represent different strategies to deal with user goals while minimizing the involved cost. Thus the next action to be selected by the DM at each system turn is the one that maximizes the corresponding heuristic function. The first two strategies (MP and MPA) deal with local decisions, i.e., they only evaluate next turn nodes and edges. Two more proposals (BP and BPA) evaluate entire paths from the actual state to a closing state in the graph. On the other hand strategies MP and BP only take into account the transition probabilities in the model whereas strategies MPA and BPA also base the decision in the amount of attributes potentially be filled.

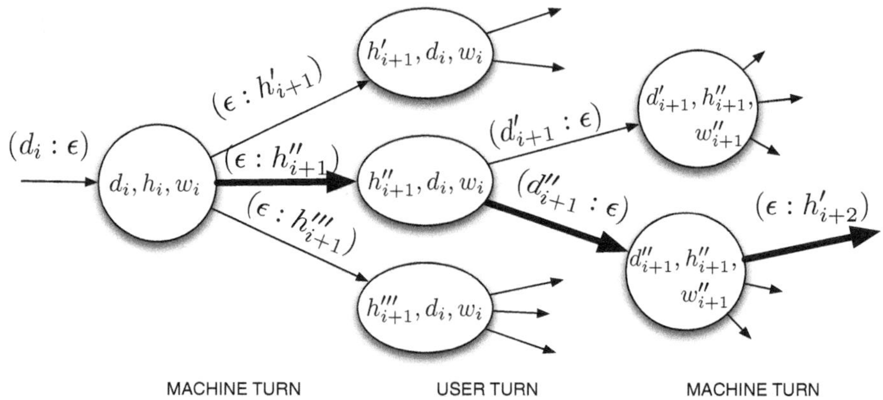

Fig. 20.2 *Bold lines* show a path θ matching some *bi-string* in a probabilistic *bi-automaton* $\mathscr{DM}$

20.3.1 Maximum Probability (MP) Strategy

This strategy just deals with Eq. (20.1). The best DM hypothesis to be selected by the DM is the one that maximizes the posterior $Pr(h|h',f)$ according to Eq. (20.2). A suboptimal approach can be considered through a two-step decoding: find first an optimal user feedback $\hat{d}$ and then, use $\hat{d}$ to decode system hypothesis $\hat{h}$ as follows:

$$\hat{d} = \arg\max_{d \in \mathscr{D}} P(f|d)P(d|h') \tag{20.6}$$

$$\hat{h} \approx \arg\max_{h \in \mathscr{H}} P(h|\hat{d}, h') \tag{20.7}$$

The dialog manager hypothesis probability distribution $P(h|d, h')$ and the user feedback probability distribution $P(f|h, d')$ have been modeled in this work by the PFSBA presented in previous section. As a consequence the search for the most likely hypothesis $\hat{h}$ in Eq. (20.7) is equivalent to choose the edge of highest transition probability at each system turn, i.e.,

$$\hat{h} = \arg\max_{h_{ij} \in \mathscr{H}(q_i)} P(q_i, (\epsilon : h_{ij}), q'_j) \tag{20.8}$$

where $q_i \in Q_{\mathscr{M}}$ is a system state labelled as $((\tilde{d}_i : \tilde{h}_i) : \tilde{w}_i)$ and $q_j \in Q_{\mathscr{U}}$ is a user turn labelled as $((\tilde{d}_j : \tilde{h}_j) : \tilde{w}_j)$ such that $P(q_i, (\epsilon : h_{ij}), q_j) > 0$, being $h_{ij} \in \mathscr{H}(q_i)$ the associated system hypotheses. This strategy has been evaluated in Ghigi et al. (2013) in a dialog generation task showing good task completion rates and good model behaviors.

20.3.2 Maximum Probability Strategy with Attributes (MPA)

We want to know the number of attributes filled as a consequence of DM decisions. Let us consider now two states of the model q_i and q_j labelled as $((\tilde{d}_i : \tilde{h}_i) : \tilde{w}_i)$ and $((\tilde{d}_j : \tilde{h}_j) : \tilde{w}_j)$. According to the model definition in Sect. 20.2 $\tilde{w}_i$ and $\tilde{w}_j$ are two sequences of symbols $w \in \Omega$ representing filled attributes in the model states i and j. We want now to define a transformation distance between $\tilde{w}_i$ and $\tilde{w}_j$ aimed at representing the number of new attributes filled in state q_j relative to the state q_i. To this end let us now consider the number $n_{del}(\tilde{w}_{ij})$ of single-symbol deletions and the number $n_{ins}(\tilde{w}_{ij})$ required to transform sequence $\tilde{w}_i$ to $\tilde{w}_j$. Let now $d_w(q_i, q_j)$ be the attribute distance between nodes q_i and q_j defined as follows:

$$d_w(q_i, q_j) = n_{ins}(\tilde{w}_{ij}) - n_{del}(\tilde{w}_{ij}) \tag{20.9}$$

Notice that we are now focussing on the number of filled attributes regardless of their associated value. Thus symbol substitutions are not considered here.

The DM has to take into account that the DM fruitful actions, like the ones aimed at consulting a database to provide the user information requirements, need the related task attributes to be previously filled. Thus this strategy is aimed at selecting a hypothesis with a high probability according to the training corpus but also at filling a high number of task attributes according to the history of the current running dialog. Notice that only user actions can change the number of filled attributes. Let $q_i \in Q_{\mathcal{M}}$ be a system state and $q_j \in Q_{\mathcal{U}}$ be a destination state $q_j \in Q_{\mathcal{U}}$ such that $P(q_i, (\epsilon : h_{ij}), q_j) > 0$ being $h_{ij} \in \mathcal{H}(q_i)$ the associated system hypothesis. The user being at state q_j can provide $|\mathcal{F}(q_j)|$ feedbacks $f_{jk} \in \mathcal{F}(q_j)$ leading to $|\mathcal{F}(q_j)|$ systems states q_{jk}. Thus the attribute distance $d_w(q_i, q_{jk})$ $k = 1, \ldots, |\mathcal{F}(q_j)|$ between node q_i and each of node q_{jk} has to be computed. Then we define a heuristic function $F(q_i, (\epsilon : h_{ij}), q_j)$ associated to each potential transition as follows:

$$F(q_i, h_{ij}, q_j) = \log P(q_i, (\epsilon : h_{ij}), q_j') + \max_{k \in |\mathcal{F}(q_j)|} d_w(q_i, q_{jk}) \tag{20.10}$$

Then the action to be taken by the system is the one that maximizes the heuristic function as follows

$$\hat{h} = \underset{h_{ij} \in \mathcal{H}(q_i)}{\arg\max}\, F(q_i, h_{ij}, q_j) \tag{20.11}$$

20.3.3 Best Path Probability (BP) Strategy

This strategy is aimed at exploring all the paths in the graph that begin in the current dialog state q_0. However only paths that lead to a closing state in the model are considered.

Let $\mathbf{z}$ be a *bi-string* over the extended alphabet $\Gamma \subseteq \Sigma^{\leq m} \times \Delta^{\leq n}$ such as $\mathbf{z} : \mathbf{z} = z_1 \dots z_{|\mathbf{z}|}$, $z_i = (\tilde{d}_i : \tilde{h}_i)$ such that the associate path $\theta_{\mathbf{z}} = (q_0, z_1, q'_1, z_2, q_2, \dots, q'_{|\mathbf{z}|-1}, z_{|\mathbf{z}|}, q_{|\mathbf{z}|})$, $q_i \in Q_{\mathcal{M}}$, $q'_i \in Q_{\mathcal{U}}$ begins in the current state of the system $q_0 \in Q_{\mathcal{M}}$. The probability $Pr_{\mathcal{DM}}(\theta_{\mathbf{z}})$ of generating $\mathbf{z}$ is calculated according to Eq. (20.5). Let now $\Theta_f(q_0)$ be the set of paths $\theta_{\mathbf{z}} \in \Theta_f(q_0)$ beginning at state q_0 and ending in a final state, i.e., $P_f(q_{|\mathbf{z}|}) = 1$. The best path $\hat{\theta}_{\mathbf{z}} \in \Theta_f(q_0)$ is the one that maximizes the normalized probability, i.e.,

$$\hat{\theta}_{\mathbf{z}} = \arg\max_{\theta_{\mathbf{z}} \in \Theta_f(q_0)} \frac{1}{|\mathbf{z}|} Pr_{\mathcal{DM}}(\theta_{\mathbf{z}}) \tag{20.12}$$

Thus the DM selects the first hypothesis $\tilde{h}_1$ defining the first element z_1 of the *bi-string* $\mathbf{z}$ associated to $\hat{\theta}_{\mathbf{z}}$, such that $z_1 = (\epsilon : \tilde{h}_1)$.

20.3.4 Best Path Probability Strategy with Attributes (BPA)

In the same way as in MPA we want now to include in the score the number of attributes filled as a consequence of DM decisions. To this end we define a heuristic function $F(\theta_{\mathbf{z}})$ associated to each $\theta_{\mathbf{z}} \in \Theta_f(q_0)$ beginning at state q_0 and ending in a final state as follows:

$$F(\theta_{\mathbf{z}}) = \sum_{i=1}^{|\mathbf{z}|} \log P(q_{i-1}, z_i, q_i) + \log P_f q_{|\mathbf{z}|} + \sum_{i=1}^{|\mathbf{z}|} (\max_{k \in |\mathcal{F}(q_j)|} d_w(q_i, q_{jk})) \tag{20.13}$$

The best path $\hat{\theta}_{\mathbf{z}} \in \Theta_f(q_0)$ is now the one that maximizes the normalized heuristic function $F(\theta_{\mathbf{z}})$, i.e.,

$$\hat{\theta}_{\mathbf{z}} = \arg\max_{\theta_{\mathbf{z}} \in \Theta_f(q_0)} \frac{1}{|\mathbf{z}|} F(\theta_{\mathbf{z}}) \tag{20.14}$$

This strategy is similar to BP, but it also takes into account the number of attributes filled at each step. Thus the DM also selects now the first hypothesis $\tilde{h}_1$ defining the first element z_1 of the *bi-string* $\mathbf{z}$ associated to $\hat{\theta}_{\mathbf{z}}$, such that $z_1 = (\epsilon : \tilde{h}_1)$.

20.4 Experiments

The four strategies described in Sect. 20.3 were evaluated over a dialog generation task from a corpus of transcribed dialogs between real users and an automatic information system.

20.4.1 Learning and Using Models to Generate Dialogs

For these experiments a DM model and a user model were estimated from Let's Go corpus (Raux et al. 2005). Let's Go is a set of spoken dialogues in English in the bus information domain. Let's Go system has been developed by Carnegie Mellon University over the Olympus–Ravenclaw framework (Bohus and Rudnicky 2009). It provides schedules and route information about the city of Pittsburgh's bus service to the general public. In this work we use a set of dialogues collected by the Let's Go Bus Information system in about 2 months, from March 2005 to April 2005. This set consists in 1840 dialogs between Ravenclaw DM and 1840 real users that include 28,141 system turns and 28,071 user turns. In order to have the Let's Go corpus labelled in terms of Dialogue Acts, we have collected the information associated to each system turn from the log files of the Ravenclaw DM, whereas the information associated to the user turn was collected from the output of the Phoenix semantic decoder. Thus we got a Σ alphabet consisting of 138 semantic symbols provided by the user and a Δ alphabet consisting of 49 dialog acts provided by the dialog manager. Additionally the attribute alphabet Ω consists of 14 attributes. A dialog example including Σ, Δ, and Ω symbols can be found at the end of Sect. 20.2.

Let us split the corpus into two subsets to train two models, one acts as a dialog manager and provides hypotheses according to the strategies defined in Sect. 20.3. The other one acts as simulated user that proposes a random user feedback in order to generate a wider variety of dialogs (Ghigi et al. 2013; Inés Torres et al. 2012). Both models have to deal with unseen events, i.e., unknown situations at training corpus. The dialog manager can provide a hypothesis $\tilde{h}_i$ that does not lead to any of the existing states created when trained from the dialog corpus. In the same way the simulated user can provide a user feedback $\tilde{f}_i$ not appearing in the training corpus, so not in the model. The generalization issue is tackled by adopting the back-off smoothing strategy proposed in Ghigi et al. (2013) for unseen events. Then a set of new dialogs were obtained from the interaction between the DM and the simulated user, as show in Fig. 20.1. For those experiments an error model simulated the ASR recognition errors. This model was trained from the dialog corpus where both the transcription of the user utterance and the output of the ASR can be found.

20.4.2 Metrics

In order to evaluate the system we have decided to use three different metrics, task completion (TC), appropriate utterance (AU), and average dialog length (ADL). The metrics used to evaluate the system are:

Task Completion (TC). Measures the success of the system in providing the user with the information requested (Danieli and Gerbino 1995). This is an automated metric and we compute it by checking if in the dialog we arrive to the point of making a query to the backend, to retrieve the information about a schedule asked by the user.

Appropriate Utterance (AU). An utterance is considered appropriate when it provides the user the required information, when it asks for additional information which is essential to respond to the user's request or when it is dealing with a repair strategy. AU evaluates whether the DM provides a coherent response at each turn according to its input (output of the ASR). We measured this metric manually, thus for each turn we check if the system answer to an user turn was appropriate.

Average Dialogue Length (ADL). The average number of turns in a dialog. A dialog that achieves the goal but has a really long length could be indication of repeated ASR errors, so the user and the system collaborate with recovering techniques in order to recover the error and the dialog gets longer.

Task Completion will give us a feedback on the global success of a single dialog. Appropriate Utterance can give us a look into the specific answer of the system. ADL can give us a feedback about the quality of a dialog.

20.4.3 Experimental Results

We carried out four sets of experiments to evaluate strategies defined in Sect. 20.3, MP, MPA, BP, and BPA, to get the best hypothesis to be provided by the DM at each interaction step. A set of 100 dialogs was generated for each strategy. TC , AU, and ADL were computed for each of the set of generated dialogs. For comparison purposes these metrics were also computed for a random set of 100 dialogs extracted from the corpus, which was conducted by Ravenclaw DM. Table 20.1 shows the results of this evaluation. This Table shows higher TC values for dialogs generated by all the proposed models than the one computed over the reference set that was managed by Ravenclaw. However the ADL value is higher in all the sets generated by the proposed models than in the reference set. Thus, the proposed formulation seems to achieve well the user goals measured in terms of TC but needs a higher number of turns to finish a dialog.

Then we compared the results obtained for each strategy shown in Table 20.1. Experimental results in Table 20.1 show that the strategies using a local decision process (MP, MPA) present significantly higher performance in terms of Task Completion. This is likely due to the unpredictable user behavior, modeled in the

Table 20.1 Task completion (TC), average dialog length (ADL), and appropriate utterance (AU) for the experiments carried out

		Local decisions		Path-based decisions	
	Ravenclaw	MP	MPA	BP	BPA
TC (%)	51.96	76.09	79.59	54.54	68.00
AU (%)	95.53	97.41	96.23	94.58	87.93
ADL	20.45	26.57	34.77	30.60	31.88

The four strategies defined in Sect. 20.3 (MP, MPA, BP, and BPA) for the DM based on Probabilistic *bi automaton* were evaluated and then compared to the Ravenclaw DM simulated user by choosing randomly the next user action. Path-based decision strategies BP and BPA select the next action as the first one in the best complete path from the current system state up to a final node. User unpredictable behavior often causes the user model to change the path in the *bi-automaton*, making worthless the selection of a best path. Furthermore we notice an increase in ADL for strategies including the attributes in the heuristic decision. This kind of heuristic tends to make the system ask for the whole set of possible attributes, also if some of them, like the bus line number, are not required. As a consequence the number of turns in generated dialogs increases. AU values are higher when local decision strategies were considered. This is due to the fact that the best path may not include the first action with the highest probability or the one with the maximum heuristic function. Tables 20.2 and 20.3 show the total number of turns (NT), the number of turns generated through a smoothed edge (NTS) and the smoothing rate (SR) representing the percentage of turns obtained through smoothing techniques of both DM and user models, when local and path-based decision strategies were considered. These tables reveal a high use of smoothed edges that underlines the importance of considering an appropriate generalization strategy. These tables also show that strategies performing local decisions seem to have a slighter lower smoothing rate percentage.

Table 20.2 Total number of turns (NT), number of turns generated through a smoothed edge (NTS) and smoothing rate (SR) representing the percentage of turns obtained through smoothing techniques for system and user turns when strategies based on local decisions were considered

	Local decision strategies					
	MP			MPA		
	Total	User	System	Total	User	System
NT	1891	944	947	2418	1207	1211
NTS	664	304	300	927	509	418
SR	35.11	32.20	38.01	38.33	42.17	34.51

Table 20.3 Number of turns generated through a smoothed edge (NTS) and smoothing rate (SR) representing the percentage of turns obtained through smoothing techniques for system and user turns when strategies based on exploring sets of paths were considered

	Path-based decision strategies					
	BP			BPA		
	Total	User	System	Total	User	System
NT	2028	994	1034	2203	1101	1102
NTS	889	493	396	849	458	391
SR	43.84	49.59	38.29	38.54	41.59	35.48

20.5 Conclusions and Future Work

In conclusion we have defined several decision making strategies over deterministic and probabilistic finite-state *bi-automaton* for dialog management. Our goal was to provide the dialog manager with the best decision at each system turn. The best system hypothesis was selected at running time according to some heuristic search aimed at achieving the user goals. Two strategies dealt with local decisions, i.e., they only evaluated the next turn nodes and edges, and obtained the best system performance on task completion. Two more proposals evaluated entire paths from the current system state to a closing state. Experimental results showed that path-based strategies that implement decisions based on possible future user actions achieved lower system performances due to unpredictability of the user behavior. Furthermore we observed a small increase in Task Completion when the heuristic function also considers the number of attributes potentially be filled by the user as a consequence of the dialog manager decisions. Ongoing work will focus on deploying a complete spoken dialog system demo and testing these strategies with real users.

Acknowledgements This work has been partially supported by the Spanish Ministry of Science under grants BES-2009-028965 and TIN2011-28169-C05-04 and by the Basque Government under grants IT685-13 and S-PE12UN061.

References

Bohus D, Rudnicky AI (2009) The Ravenclaw dialog management framework: architecture and systems. Comput Speech Lang 23:332–361

Churcher GE, Atwell ES, Souter C (1997) Dialogue management systems: a survey and overview. Research report Series-University of Leeds, School of Computer Studies, LU SCS RR

Danieli M, Gerbino E (1995) Metrics for evaluating dialogue strategies in a spoken language system. In: Proceedings of the 1995 AAAI spring symposium on empirical methods in discourse interpretation and generation, vol 16, pp 34–39

Ghigi F, Inés Torres M, Justo R, Benedí J-M (2013) Evaluating spoken dialogue models under the interactive pattern recognition framework. In: INTERSPEECH, pp 480–484

Hurtado LF, Planells J, Segarra E, Sanchis E, Griol D (2010) A stochastic finite-state transducer approach to spoken dialog management. In: INTERSPEECH, pp 3002–3005

Inés Torres M (2013) Stochastic bi-languages to model dialogs. In: Proceedings of international workshop on finite state methods and natural language processing, p 9

Inés Torres M, Benedí JM, Justo R, Ghigi F (2012) Modeling spoken dialog systems under the interactive pattern recognition framework. In: SSPR&SPR. Lecture notes on computer science, pp 519–528

Jurčíček F, Thomson B, Young S (2012) Reinforcement learning for parameter estimation in statistical spoken dialogue systems. Comput Speech Lang 26(3):168–192

Lee C, Jung S, Eun J, Jeong M, Lee GG (2006) A situation-based dialogue management using dialogue examples. In: Proceedings of the IEEE international conference on acoustics, speech, and signal processing (ICASSP), pp 69–72

Raux A, Langner B, Bohus D, Black AW, Eskenazi M (2005) Let's go public! taking a spoken dialog system to the real world. In: Proceedings of Interspeech
Seneff S, Polifroni J (2000) Dialogue management in the mercury flight reservation system. In: Proceedings of the 2000 ANLP/NAACL workshop on conversational systems, Stroudsburg, vol 3, pp 11–16
Thomson B, Yu K, Keizer S, Gasic M, Jurcicek F, Mairesse F, Young S (2010) Bayesian dialogue system for the let's go spoken dialogue challenge. In: 2010 IEEE workshop on spoken language technology workshop (SLT), IEEE, pp 460–465
Toselli AH, Vidal E, Casacuberta F (eds) (2011) Multimodal interactive pattern recognition and applications. Springer, London
Williams JD, Young S (2007) Partially observable Markov decision processes for spoken dialog systems. Comput Speech Lang 21:393–422

Chapter 21
Integration of Word and Semantic Features for Theme Identification in Telephone Conversations

Yannick Estève, Mohamed Bouallegue, Carole Lailler, Mohamed Morchid, Richard Dufour, Georges Linarès, Driss Matrouf, and Renato De Mori

Abstract The paper describes a research about the possibility of integrating different types of word and semantic features for automatically identifying themes of real-life telephone conversations in a customer care service (CCS). Features are all the words of the application vocabulary, the probabilities obtained with latent Dirichlet allocation (LDA) of selected discriminative words and semantic features obtained with a limited human supervision of words and patterns expressing entities and relations of the application ontology. A deep neural network (DNN) is proposed for integrating these features. Experimental results on manual and automatic conversation transcriptions are presented showing the effective contribution of the integration. The results show how to automatically select a large subset of the test corpus with high precision and recall, making it possible to automatically obtain theme mention proportions in different time periods.

Mohamed Bouallegue thanks the ANR agency for funding through the CHIST-ERA ERA-Net JOKER project.

Carole Lailler thanks European Commission for funding through the EUMSSI Project, number 611057, call FP7-ICT-2013-10.

Y. Estève (✉) • M. Bouallegue • C. Lailler
LIUM, University of Le Mans, Le Mans, France
e-mail: yannick.esteve@univ-lemans.fr; mohamed.bouallegue@lium.univ-lemans.fr; carole.lailler@lium.univ-lemans.fr

M. Morchid • R. Dufour • G. Linarès • D. Matrouf
LIA, University of Avignon, Avignon, France
e-mail: mohamed.morchid@univ-avignon.fr; richard.dufour@univ-avignon.fr; georges.linares@univ-avignon.fr; driss.matrouf@univ-avignon.fr

R. De Mori
LIA, University of Avignon, Avignon, France

McGill University, Montreal, QC, Canada
e-mail: rdemori@cs.mcgill.ca

G. Geunbae Lee et al. (eds.), *Natural Language Dialog Systems and Intelligent Assistants*, DOI 10.1007/978-3-319-19291-8_21

Keywords Theme identification • Human–human spoken conversation • Deep neural network

21.1 Introduction

A growing research interest has been observed in the automatic analysis of human/human spoken conversations as reviewed in Tur and De Mori (2011) and Tur and Hakkani-Tür (2011). A scientifically interesting and practically important component of this research is topic identification for which an ample review of the state of the art can be found in Hazen (2011). In spite of the relevant progress achieved so far, it is difficult to reliably identify multiple topics in real-life telephone conversations between casual speakers in unpredictable acoustic environments. Of particular interest are call-centre conversations in which customers discuss problems in specific domains with an advisor. This is the case of the application considered in this paper. The purpose of the application is to collect statistics about the problems discussed in the customer care service (CCS) of the ratp Paris transportation system. Statistics are obtained by analysing real-world human/human telephone conversations in which an agent attempts to solve a customer problem. Proportions of problem themes are used for monitoring user establishing priorities of problem solving interventions. Application relevant information for the task is described in the application requirements. Themes are the most general entitles of the application ontology outlined in the documentation. Agents follow a pre-defined protocol to propose solutions to user problems about the transportation system and its services. An automatic classification of conversation themes is motivated by the fact that, due to time constraints, agents cannot adequately take note of the discussed themes. A fully automatic system for theme identification must include an automatic speech recognition (ASR) module for obtaining automatic transcriptions of the conversations. The acoustic environment on these conversations is unpredictable with a large variety of noise types and intensity. Customers may not be native French speakers and conversations may exhibit frequent disfluencies. The agent may call another service for gathering information. This may cause the introduction of different types of non-speech sounds that have to be identified and discarded. For all these reasons, the word error rate (WER) of the asr system is highly variable and can be very high.

Popular features for topic identification are reviewed in Hazen (2011). Concise representations of document contents have been proposed using features obtained with latent semantic analysis (LSA) (Deerwester et al. 1990), probabilistic latent semantic analysis (pLSA) (Li), and latent Dirichlet allocation (LDA) (Blei et al. 2003). Among them, LDA features provide rich representations in latent spaces with a limited number of dimensions. Recently, in Morchid et al. (2014) a detailed analysis has been reported in terms of theme classification accuracy in spoken conversations by varying the word vocabulary size and the number of hidden topics. This suggested performing a detailed analysis of classification accuracy by fine-grained variations of the number of hidden topics (Morchid et al. 2014), and the value α of the LDA hyperparameter (Morchid et al. 2014). As a large variation of the classification accuracy was observed, it was proposed to compose each feature set obtained with a specific hidden space size and value of α in a single vector called c-vector. The best improvement over other types of features was observed for a c-vector whose elements are unigram probabilities of a limited set of discriminative words.

In the application considered in this paper, the conversions to be analysed are made available by relatively small sets collected in different time periods. Relevant discriminative words may change in time even if most of them belong to the same semantic category defined in the application ontology. An example is a bus line whose itinerary has been temporarily modified. The names of some streets are likely to be frequently mentioned in that time period by customers inquiring about the itinerary. These specific names are unlikely to have been selected as discriminative words even if their co-presence with other words is very useful to characterize a traffic state. A new approach is proposed in this paper to embed different types of latent features, some of them representing words of the entire vocabulary as proposed in Sarikaya et al. (2014) for call routing, some other representing expressions of fragments of the application ontology and some others being LDA features. Learning is supervised since conversations are annotated in terms of themes, while some features are obtained with minor supervision of a list of automatically proposed candidates. Completely unsupervised approaches for other semantic interpretation tasks are proposed in Chen et al. (2014) and Cuayáhuitl et al. (2014).

21.2 Features Used for Theme Identification

The corpus of the application considered in this paper is annotated with eight conversation themes: *problems of itinerary, lost and found, time schedules, transportation cards, state of the traffic, fares, infractions* and *special offers*. Three types of features are considered for theme identification. They are the words of the application vocabulary V_W, a set S of labels for application concepts and a conversation summary represented by a c-vector whose elements are unigram probabilities of words belonging to a reduced vocabulary $V_S \subset V$ of theme discriminative words.

21.2.1 Word Features

All the words of the application vocabulary V_W are considered as features for theme identification. For each conversation, a vector W of binary values is built in which each element corresponds to a word in V_W and its value is set to 1 only if the corresponding word is present in the conversation

A discriminative word vocabulary $V_D \subset V_W$ is formed as described in Morchid et al. (2014) with the top 116 words of V_W ranked with the product of term frequency

(TF), inverse document frequency (IDF) and word purity in the themes. Unigram probabilities of discriminative words are computed in an r-dimensional hidden space using LDA as follows:

$$P_r(w_i|d) = \sum_{n=1}^{N_r} P(w_i|z_n^r)P(z_n^r|d) \tag{21.1}$$

where N_r is the number of hidden spaces, $P(w_i|z_n^r)$ is the probability of word w_i in the n-th hidden topic of the r-th hidden space and $P(z_n^r|d)$ is the probability of the n-th hidden topic in the d-th conversation.

Let $x^r(d) = P_r(w|d)$ be the vector having probabilities $P_r(w_i|d)$ as elements. These probabilities are estimated using hidden topic spaces z_n^r that are different for each hidden space considered and thus are not the themes of the conversations. All vectors $x^r(d)$ are then integrated into a unique C *vector* obtained with Joint Factor Analysis as described in Morchid et al. (2014).

21.2.2 Semantic Features

Conversations of the train set labelled with only one theme are selected. Using them, a unigram language model (LM) is obtained for each theme and for the ensemble of all the conversations.

Let $\Im_k$ be the set of conversations of theme τ_k. For each conversation theme, a set of words is selected using the approach described in Carpineto et al. (2001).

Let $P_k(w)$ represent the LM probability distribution estimated with the data in $\Im_k$ and $P_g(w)$ represent the probability distribution estimated with the data in $\Im_g = \cup_{k=1}^{K} \Im_k$, where K is the number of conversation themes.

The two distributions $P_k(w)$ and $P_g(w)$ diverge and a measure of their divergence is the *Kullback–Leibler* divergence (KLD) measure:

$$\text{KLD}[P_k(w), P_g(w)] = \sum_{w \in \Im_g} P_k(w) \log \frac{P_k(w)}{P_g(w)} \tag{21.2}$$

It has been shown in Carpineto et al. (2001) that, when comparing word unigram distributions, the addends that mostly contribute with a positive value to the summation in $\text{KLD}[P_k(w), P_g(w)]$ are useful features for performing relevance feedback in information retrieval. The same approach is applied to the train set for making a list of words for each conversation theme. Another application of an approach of this type can be found in Wu et al. (2010). A human expert analyses the words of each list starting from the top of the list. Words that express facts and other concepts specific to each theme of the application ontology are selected and labelled with concepts of the application ontology. Generalizations are performed by associating the same concept to words not observed in the train set but belonging to the same class record of the application database.

Let V_S be the vocabulary of these concept labels. For each conversation, a vector S of binary values is built in which each element corresponds to a concept in V_S and its value is set to 1 only if the corresponding concept is present in the conversation. Features embedding the co-presence of word and concept features are automatically learned in the approach proposed in this paper. For example, the co-presence of a location concept with at least two different locations words is expected to be a useful feature for the theme itinerary. Other dependencies expressing the co-presence of words and concepts are expected to be useful expressions of application relevant distant semantic relations. Simple patterns involving words and concepts expressing local semantic relations are also manually derived with a minor human effort. The corresponding meanings are also represented by elements of vector S.

21.3 A Deep Neural Network Architecture for Theme Identification

A deep neural network (DNN) architecture is proposed for integrating word features represented by vector W, semantic features represented by vector S and conversation summaries represented by the C vector.

This architecture should be simple enough to allow effective training of its parameters to obtain features that correspond to requirements inspired by the application ontology. The most important requirement is to capture sufficient concept relations between concepts characterizing each theme. The second requirement is to ensure coherence between specific dependencies expressed by hidden features and a concise global representation of a conversation expressed by LDA features.

In order to capture some dependencies between words, concepts expressed by words and semantic entities expressed by short distance patterns vectors W and S are concatenated and encoded into a vector $X = SW$ by the equation:

$$h_1(X) = f(U \times X + b_1) \tag{21.3}$$

The elements of matrix U and vector b_1 are estimated by a multi-layer perceptron having vector X as input, vector $h_1(X)$ computed by the hidden layer and a vector V_K of $K = 8$ output nodes corresponding to the 8 themes of the application.

The vector $h_1(X)$ is then concatenated with an embedding:

$$h_2(C) = f(B \times C + b_2) \tag{21.4}$$

of vector C to obtain a vector:

$$Q = h_1(X)h_2(C) \tag{21.5}$$

A vector $h_3(Q)$ is obtained at a third hidden layer with the relation:

$$h_3(Q) = f(G \times Q + b_3) \tag{21.6}$$

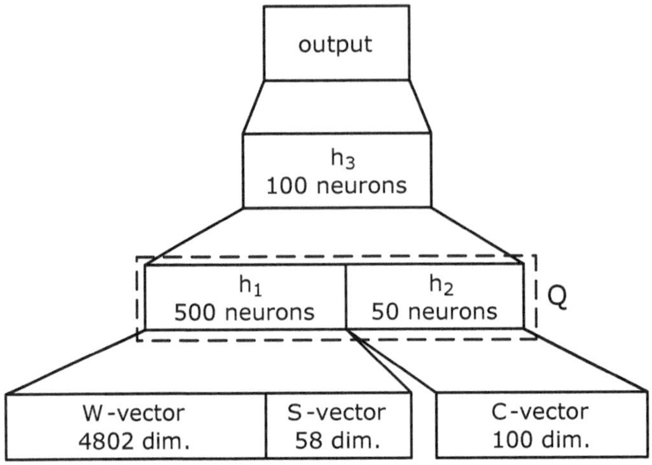

Fig. 21.1 DNN architecture for theme identification

The DNN output values are computed as

$$V_k = f(L \times h_3(Q) + b_4) \tag{21.7}$$

Eventually only the elements of matrices B and Q are estimated with back propagation while the values of U are kept fixed. The reason is that in this way the number of parameters to be estimated is kept relatively small to avoid over fitting due to the limited size of the train set.

The resulting DNN architecture is depicted in Fig. 21.1. Substructures of this architecture have also been used to compare results obtained with subsets of the considered features.

21.4 Experimental Set-Up

The corpus of the DECODA project (Béchet et al. 2012) has been used for the theme identification experiments described in this section. This corpus is composed of 1067 telephone conversations from the call centre of the public transportation service in Paris. The corpus is split into a train set (740 dialogues) and a test set (327 dialogues). Conversations have been manually transcribed and labelled with one theme label corresponding to the principal concern mentioned by the customer. A portion of the train set (175 dialogues) is also used as a development set for selecting the dimension of the hidden topic spaces. All hidden spaces were obtained with the manual transcriptions of the train set. The number of turns in a conversation and the number of words in a turn are highly variable. The majority of the conversations have more than ten turns. The turns of the customer tend to be longer (>20 words) than those of the agent and are more likely to contain out of vocabulary words that are often irrelevant for the task.

The ASR system used for the experiment is the LIA-speeral system (Linarès et al. 2007) with 230,000 Gaussians in the triphone acoustic models. Model parameters were estimated with maximum a-posteriori probability (MAP) adaptation of 150 h of speech in telephone bandwidth with the data of the train set. The vocabulary contains 5782 words. A 3-gram language model (LM) was obtained by adapting with the transcriptions of the train set a basic LM. An initial set of experiments were performed with this system resulting with an overall WER on the test set of 58 % (53 % for agents and 63 % for users). These high error rates are mainly due to speech disfluencies and to adverse acoustic environments for some dialogues when, for example, users are calling from train stations or noisy streets with mobile phones. Furthermore, the signal of some sentences is saturated or of low intensity due to the distance between speakers and phones.

Experiments were performed with different types of inputs and components of the network whose scheme is shown in Fig. 21.1. For the sake of comparison, a set of experiments were performed with the manual transcriptions (TRS) of the conversations using simple multi-layer perceptrons (MLP) with one hidden layer, an output layer with K nodes and fed by different input vectors. The results for different architectures are reported in Table 21.1 for the development (DEV) and the test (TEST) sets. The confidence interval is ± 3.69.

The results of Table 21.1 show a superiority by using as input the word vector W rather than the vector S of semantic labels and a further improvement by concatenating vectors W and S at the input.

The same type of experiment was performed using ASR transcriptions (ASR). The results are reported in Table 21.2. An improvement is again observed by concatenating W with S at the input of an MLP network. Minor improvements are observed by concatenating the C-vector since it represents another abstraction compared with the embedding of the other features. Probably both abstractions capture the same large proportion of semantic content. Nonetheless, the improvement is inside the confidence interval, suggesting to consider additional input features in the architecture with the structure shown in Fig. 21.1, indicated as DNN in Table 21.2.

Table 21.1 Percent accuracies for MLP architectures, for the development (DEV) and the test (TEST) sets

Input	Train/test corpus	DEV	TEST
W	TRS/TRS	86.9	83.2
S	TRS/TRS	82.9	78.9
W+S	TRS/TRS	89.7	85.9

The data of the train and test sets are manual transcriptions (TRS)

Table 21.2 Percent accuracies for different architectures for the test (TEST) set

Input	Train/test corpus	Architecture	TEST
W	ASR/ASR	MLP	79.5
W+S	ASR/ASR	MLP	82.3
W+S+C	ASR/ASR	DNN	82.9

The data of the train and test sets are ASR transcriptions (ASR)

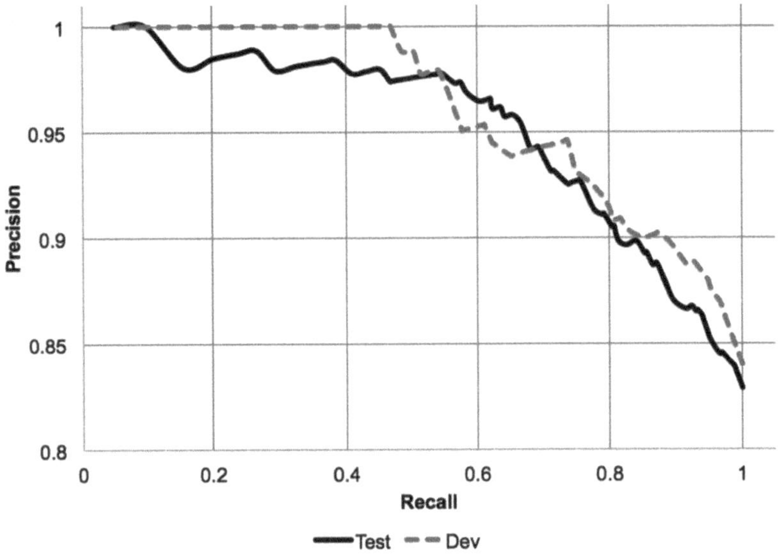

Fig. 21.2 Precision–recall results for the test set

The results are promising, but more interesting is the precision–recall relation shown in Fig. 21.2. It has been obtained by selecting conversations based on the posterior probability $P_{k1}(d)$ of the theme t_1 ranked first for conversation d. Posterior probabilities of theme hypotheses for conversation d are computed with the *softmax* function applied to the outputs of DNN fed by features of d. The curve shows that a precision of 90 % can be achieved with 84 % recall, making it possible to obtain practically useful conversation survey proportions with a small rejection of samples that could be manually annotated with a limited effort. These results compare favourably with the best results obtained so far with the same corpus (Morchid et al. 2014) where the 90 % precision was obtained with 78 % recall.

21.5 Conclusion

A DNN architecture has been proposed for theme identification in human/human conversations by integrating different word and semantic features. With the proposed network high precisions can be obtained by rejecting a small proportion of conversations classified with high equivocation. As some of the semantic features used for theme identification are descriptions of basic facts characterizing a theme, it will be possible in future work to automatically verify the mentions of basic facts for an automatically identified theme. An absence of fact mention is a clue for selecting the conversation as an informative example to be manually analysed for discovering and generalizing possible new mentions of basic facts for one or more themes discussed in the selected conversation.

References

Béchet F, Maza B, Bigouroux N, Bazillon T, El-Bèze M, De Mori R, Arbillot E (2012) Decoda: a call-centre human-human spoken conversation corpus. In: Proceeding of LREC'12

Blei DM, Ng AY, Jordan MI (2003) Latent Dirichlet allocation. J Mach Learn Res 3:993–1022

Carpineto C, De Mori R, Romano G, Bigi B (2001) An information-theoretic approach to automatic query expansion. ACM Trans Inf Syst 19(1):1–27

Chen Y-N, Wang WY, Rudnicky AI (2014) Leveraging frame semantics and distributional semantics for unsupervised semantic slot induction in spoken dialogue systems. In: IEEE spoken language technology workshop (SLT 2014), South Lake Tahoe, California and Nevada

Cuayáhuitl H, Dethlefs N, Hastie H, Liu X (2014) Training a statistical surface realiser from automatic slot labelling. In: IEEE spoken language technology workshop (SLT 2014), South Lake Tahoe, California and Nevada

Deerwester S, Dumais ST, Furnas GW, Landauer TK, Harshman R (1990) Indexing by latent semantic analysis. J Am Soc Inf Sci 41(6):391–407

Hazen TJ (2011) MCE training techniques for topic identification of spoken audio documents. IEEE Trans Audio Speech Lang Process 19(8):2451–2460

Linarès G, Nocéra P, Massonie D, Matrouf D (2007) The LIA speech recognition system: from 10xRT to 1xRT. In: Proceedings of the 10th international conference on text, speech and dialogue. Springer, Berlin, pp 302–308

Morchid M, Dufour R, Bousquet P-M, Bouallegue M, Linarès G, De Mori R (2014) Improving dialogue classification using a topic space representation and a gaussian classifier based on the decision rule. In: Proceedings of ICASSP

Morchid M, Bouallegue M, Dufour R, Linarès G, Matrouf D, De Mori R (2014) An i-vector based approach to compact multi-granularity topic spaces representation of textual documents. In: The 2014 conference on empirical methods on natural language processing (EMNLP), SIGDAT

Morchid M, Bouallegue M, Dufour R, Linarès G, Matrouf D, De Mori R (2014) I-vector based representation of highly imperfect automatic transcriptions. In: Conference of the international speech communication association (INTERSPEECH) 2014, ISCA

Morchid M, Dufour R, Bouallegue M, Linarès G, De Mori R (2014) Theme identification in human-human conversations with features from specific speaker type hidden spaces. In: Fifteenth annual conference of the international speech communication association

Sarikaya R, Hinton GE, Deoras A (2014) Application of deep belief networks for natural language understanding. IEEE/ACM Trans Audio Speech Lang Process 22(4):778–784

Tur G, De Mori R (2011) Spoken language understanding: systems for extracting semantic information from speech. Wiley, New York

Tur G, Hakkani-Tür D (2011) Human/human conversation understanding. In: Spoken language understanding: systems for extracting semantic information from speech. Wiley, New York, pp 225–255

Wu MS, Lee HS, Wang HM (2010) Exploiting semantic associative information in topic modeling. In: Proceedings of the IEEE workshop on spoken language technology (SLT 2010), pp 384–388

Chapter 22
CLARA: A Multifunctional Virtual Agent for Conference Support and Touristic Information

Luis Fernando D'Haro, Seokhwan Kim, Kheng Hui Yeo, Ridong Jiang, Andreea I. Niculescu, Rafael E. Banchs, and Haizhou Li

Abstract In this paper we present a multifunctional conversational agent which combines natural language search capabilities for two different domain applications: a conference information system and local tourist guide. The paper describes the corpora, architecture, algorithm, and the mobile application created to interact with the users. Finally, some results obtained when using the proposed system in the context of an international scientific conference held in Singapore in September 2014 with more than 1200 assistants are provided.

Keywords Conversational agents • Chatbot • Information retrieval • Q&A

22.1 Introduction

Nowadays, there is an increasing interest in using conversational agents for both web and mobile applications since they allow users to quickly find corporate or product information while, at the same time, engage users by providing them with relevant notifications about new products, offering recommendations, or simply by being able to handle user complaints or feedback (i.e. chatbot capabilities).

There are several examples of conversational agents used for different domains in the literature, such as health-care (Beveridge and Fox 2006), weather forecast (Maragoudakis 2007), tutoring (Johnson and Valente 2008), tourism (Niculescu et al. 2014), etc. Probably, the most popular applications are Apple's Siri

L.F. D'Haro (✉) • S. Kim • K.H. Yeo • R. Jiang • A.I. Niculescu • R.E. Banchs • H. Li
Human Language Technology Department, Institute for Infocomm Research, 1 Fusionopolis Way, #21-01 Connexis (South Tower), Singapore 138632, Singapore
e-mail: luisdhe@i2r.a-star.edu.sg

G. Geunbae Lee et al. (eds.), *Natural Language Dialog Systems and Intelligent Assistants*, DOI 10.1007/978-3-319-19291-8_22

(Bellegarda 2014), Google Now,[1] and Cortana.[2] These systems are able to provide information for multiple tasks and domains including making appointments, sending text messages, providing weather or transportation information, or searching the web. However, up to the best of our knowledge, there is not any conversational agent specifically designed for conferences, which at the same time could handle local information about the place where the conference is held, which certainly is very useful for first time visitors.

The paper is organized as follows: in Sect. 22.2, we describe the conference and tourist datasets used. Section 22.3 presents the system architecture explaining each module and algorithms for handling different types of searches. Lastly, in Sect. 22.4, we show some system usage statistics collected when used in a real conference.

22.2 Data Description

Since our goal was to deploy this application within the context of a real and large scientific conference, we decided to launch it during the 15th conference of the International Speech Communication Association[3] (Interspeech), which is an annual conference where more than 1200 attendees from 46 different countries meet together to discuss and share information about speech-related technologies. The conference was held in Singapore from 14 to 18 of September 2014 and included 93 different sessions (see Table 22.1).

The information concerning the conference was extracted from the official proceedings. In this case, we used only the titles, abstracts, and sessions but not the paper content itself. The main reason for not using the full text was to adjust

Table 22.1 Main statistics related to conference and tourism information domains

Domain	Concept	Details
Conference	No. paper titles	633
Conference	No. sessions	93
Conference	No. keywords extracted from titles, abstracts and sessions names	7519
Conference	No. abbreviations and scientific terms	1276
Conference	No. authors	1695
Conference	No. countries from the assistants	46
Tourism	No. of different kind of restaurant-related places	28
Tourism	No. of places to search for restaurant-related information	120
Tourism	No. of different kind of food to search for	81
Tourism	No. of main touristic attractions	35

[1] http://www.google.com/landing/now/.

[2] http://www.windowsphone.com/en-sg/how-to/wp8/cortana/meet-cortana.

[3] http://www.interspeech2014.org.

better our system to the type of information people usually search for, i.e. by topic, nationality or affiliations, specific authors, or some hot-topic technology, rather than searching specific info about how an algorithm works or the final results of a system, which are detailed in the paper.

On the other hand, and considering that this conference was held in Singapore for the first time, we decided to provide some tourism information to visitors too. Taking into account that we have already deployed a tourist conversational agent (Niculescu et al. 2014) and a restaurant recommendation system (Kim and Banchs 2014), we decided to include these two functionalities into the new system, allowing the attendees to find useful information about sightseeing, transportation, shopping centres, food and beverage, as well as some general but important information about Singapore (see Table 22.1).

22.3 Architecture Description

The system architecture (Fig. 22.1) has three main components: (1) the client system implemented in a mobile application, (2) a websocket server which runs the service and internally communicates with the orchestration and searching modules, and (3) the different resources (e.g. databases, dictionaries, and models) used to provide the information to the users and to enhance the system capabilities.

In more detail, the user poses a query using the graphical interface available on the mobile. With this information, the system creates a JSON message consisting of the query, the domain (conference or tourist), and GPS coordinates, if the user allowed for sharing them. Then, the server calls the orchestration module which, depending on the specified domain, follows the following process:

First, the system searches in the index for generic questions or greetings like: *what can you do?*, *What is your name?*, or *how can I start using you?* The search is done by retrieving the most similar examples in the index to the input query with a very high threshold to guarantee a high precision in the answer selection process.

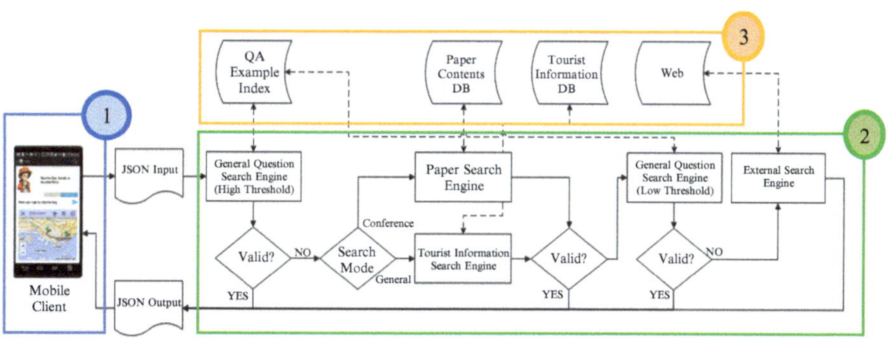

Fig. 22.1 System architecture

In case there is not an answer above the predefined threshold, the orchestration module calls either the conference search engine or the tourist search engine depending on the specific domain under consideration.

In the case of paper searches (conference domain), the system allows users to search by authors, affiliations, countries, titles, conference sessions and events, as well as general queries about conference facilities. Here, the system recognizes conference domain entities from the input by means of a fuzzy search algorithm; the algorithm is robust to a certain degree to input misspellings (this is particularly important for the case of authors' names, where a high incidence of misspellings is expected). Then, the extracted entities are expanded with knowledge bases, including the information of keywords, abbreviations, and synonyms. For each input, an SQL query is generated based on this semantic representation and it is used for searching papers, authors, or session information in the conference database.

For example, the system could capture the concept *Microsoft* as an affiliation and *speaker recognition* as a technical term from the user input *show me papers from MS on speaker recognition*, and then an SQL query with these constraints in the *where* clause would be prepared.

For the case of tourist information (tourism domain), the system is able to search for maps, restaurants, sightseeing locations, as well as for local and general information (history, currency, exchange rates, laws, etc.). Here, the module includes a text classifier that is able to detect the topic of search. Depending on this classification, the system parses the query in order to provide meaningful answers.

For instance, in case the user is doing a search like *show restaurants serving cheap Thai food in the nearby*, the module extracts the following semantic information: (pricerange, *cheap*), (type_of_food, *Thai*), and (location, *nearby*). The same fuzzy matching algorithm as in the paper search engine is used for this parsing process. With this information, the system first disambiguates the location by replacing the word *nearby* with the GPS coordinates and searches for the district where the user is located. Next, the system creates a structured URL on a well-known restaurant website in Singapore, which will be used later to display the results on the mobile app. In addition, the system is also able to handle queries like: *show me the nearest subway station* or *a good shopping center in downtown*. In these cases, the system creates maps or routes by using Google Maps API.[4]

Following with the algorithm, if there is still not an answer, the system attempts a new search in the index for generic queries, but using a more relaxed threshold. In the last instance, the system tries to search on internet by retrieving the result of a search using Wolfram Alpha search engine.[5]

Finally, the system sends back the answer to the mobile app by generating a JSON message containing the following information. (a) The best agent to show the information (in case there is a change with respect to the original agent requested by the user, the mobile application switches to the new agent providing this way a

[4]https://developers.google.com/maps/.

[5]http://www.wolframalpha.com/.

feedback to the user regarding the agent used to answer the question). (b) Feedback information displayed in a textbox near the avatar. (c) Type of information to show: this informs the mobile if the information to display is a map, an external website, directly a HTML content, a list of papers/authors/sessions, or simply a chat answer. (d) The URL to be displayed: this way it is possible to show maps, websites, restaurants information, or pictures from sightseeing points.

22.4 Deployment and Results

In order to implement the final user interface, a free mobile app, available at the Google Play and Apple's App stores, was deployed in collaboration with researchers from the Quality and Usability Lab of Telekom Innovation Laboratories (TU Berlin). The full version of the app included several capabilities such as: (a) basic search for papers, authors and sessions, (b) conference schedule, (c) the possibility of adding events to the user's personal calendar, (d) conference venue maps, (e) direct access to the conference website, and (f) a paper recommendation system.

In addition, the mobile application included a tab where the users could find our conversational agent (Fig. 22.2). The agent screen was divided into three sections.

Fig. 22.2 Aspect of the agent interface showing the three main areas. In the *left*, an example of search for tourist domain, in the *right* an example of search for papers

Table 22.2 Usage statistics during the conference

Concept	Counts
No. total queries	2360
No. different queries	1628
No. total different users	215
No. queries related with tourism/restaurants/maps	222
No. queries related with papers	480
No. chat interactions	1598

(1) The avatar and a feedback textbox to display answers or summaries of the results. In addition, this section included a switch button for the user to select the type of information they were looking for. In case of looking for conference information, they could use the agent wearing a formal suit (conference domain system). Alternatively, in case of looking for touristic information, they should use the agent wearing the informal suit (tourism domain system). (2) The input box and submit button for sending queries to the system. (3) A multipurpose view area, where the information retrieved by the system is displayed (e.g. websites, list of clickable papers, maps, restaurants, history of previous searches and results, etc.).

Finally, Table 22.2 shows some of the usage statistics we collected during the conference. We can see that, in general, people used more the application for chatting with the system and searching for touristic information than for searching for conference information. The low usage of this latter information system can be probably due to the presence of the built-in search capabilities offered by the app, the recommender mechanism which also offered push notifications to the user, and some chat game sponsored by the conference organizers for people using the mobile app. On the other hand, we found that the number of queries the system could not answer was quite high (about 50 % of the queries), but most of these (around 75 %) corresponded to out-of-domain queries (i.e. mainly chat interactions), which the system could not detect as part of the chat and tried to get an answer from the index or external websites therefore being unable to process them.

22.5 Conclusions and Future Work

In this paper we have presented a conversational agent that is able to answer to natural language questions formulated in the context of a scientific conference, as well as local information. In the conference domain, the system supports paper search by author names, topic, affiliation, and country, while in the tourism domain the system is able to provide information related to touristic places, restaurants, and transportation. In addition, the system can chat with the users. Finally, the deployed system was tested during a large conference with satisfactory results.

As future work we want to implement a mechanism to allow system administrators for conducting quick updates and sending notifications to the users. With this mechanism, if the system is not able to provide an answer, new information can be updated into the system and the user will receive a push notification on the mobile when an answer to his/her question is available.

Acknowledgements We want to thank people from the Institute for Infocomm Research (I2R), Nanyang Technological University of Singapore (NTU), and the Insterspeech organizing committee for their support on deploying and testing the system. We also want to thank Dr. Jochen Walter Ehnes (I2R), Nicholas de Laczkovich, and Tilo Westermann from the Quality & Usability Lab of Telekom Innovation Laboratories (TU Berlin) for their work to integrate the virtual agent on the mobile app.

References

Bellegarda J (2014) Spoken language understanding for natural interaction: the Siri experience. In: Natural interaction with robots, knowbots and smartphones: putting spoken dialog systems into practice. Springer, New York, pp 3–14. doi:10.1007/978-1-4614-8280-2_1

Beveridge M, Fox J (2006) Automatic generation of spoken dialogue from medical plans and ontologies. Biomed Inform 39(5):482–499. doi:10.1016/j.jbi.2005.12.008

Johnson WL, Valente A (2008) Tactical language and culture training systems: using artificial intelligence to teach foreign languages and cultures. In: Proceedings of IAAI, pp 1632–1639. doi:10.1609/aimag.v30i2.2240

Kim S, Banchs RE (2014) R-cube: a dialogue agent for restaurant recommendation and reservation. In: Proceedings of APSIPA, special session on chatbots and dialogue agents, Siem Reap, Cambodia, pp 1–6. doi:10.1109/APSIPA.2014.7041732

Maragoudakis M (2007) MeteoBayes: effective plan recognition in a weather dialogue system. IEEE Intell Syst 22(1):66–77. doi:10.1109/MIS.2007.14

Niculescu AI, Jiang R, Kim S, Yeo K-H, D'Haro LF, Niswar A, Banchs RE (2014) SARA: Singapore's Automated Responsive Assistant, a multimodal dialogue system for touristic information. In: Proceedings of MobiWIS, Barcelona, Spain, pp 153–164. doi:10.1007/978-3-319-10359-4_13

Chapter 23
Multi-Source Hybrid Question Answering System

Seonyeong Park, Hyosup Shim, Sangdo Han, Byungsoo Kim, and Gary Geunbae Lee

Abstract In this demonstration, we present a multi-source hybrid Question Answering (QA) system. Our system consists of four sub-systems: (1) a knowledgebase based QA, (2) an information retrieval based QA, (3) a keyword QA and (4) an information-extraction to construct our own knowledgebase from web texts. With these sub-systems, we can query three types of information sources: curated knowledgebases, automatically constructed knowledgebases and wiki texts.

Keywords knowledgebase based QA • Information retrieval based QA • Information extraction • Keyword QA • knowledgebase construction

23.1 Introduction

Various approaches to QA based on knowledge base (KB) have been proposed. QA is evolving from systems based on information retrieval (IR) to systems based on KBs. QA based on KBs gives very high precision, but requires curated KBs; but these KBs cannot cover all the information that web text can convey. To solve this limitation, multiple information sources other than curated KBs are needed. In this demo, we present a hybrid QA system (Fig. 23.1) that uses multiple information sources: a curated KB, an automatically constructed KB and web text.

S. Park (✉) • H. Shim • S. Han • B. Kim • G.G. Lee
Pohang University of Science and Technology, Pohang, Republic of Korea
e-mail: sypark322@postech.ac.kr; hyosupshim@postech.ac.kr; hansd@postech.ac.kr; bsmail90@postech.ac.kr; gblee@postech.ac.kr

G. Geunbae Lee et al. (eds.), *Natural Language Dialog Systems and Intelligent Assistants*, DOI 10.1007/978-3-319-19291-8_23

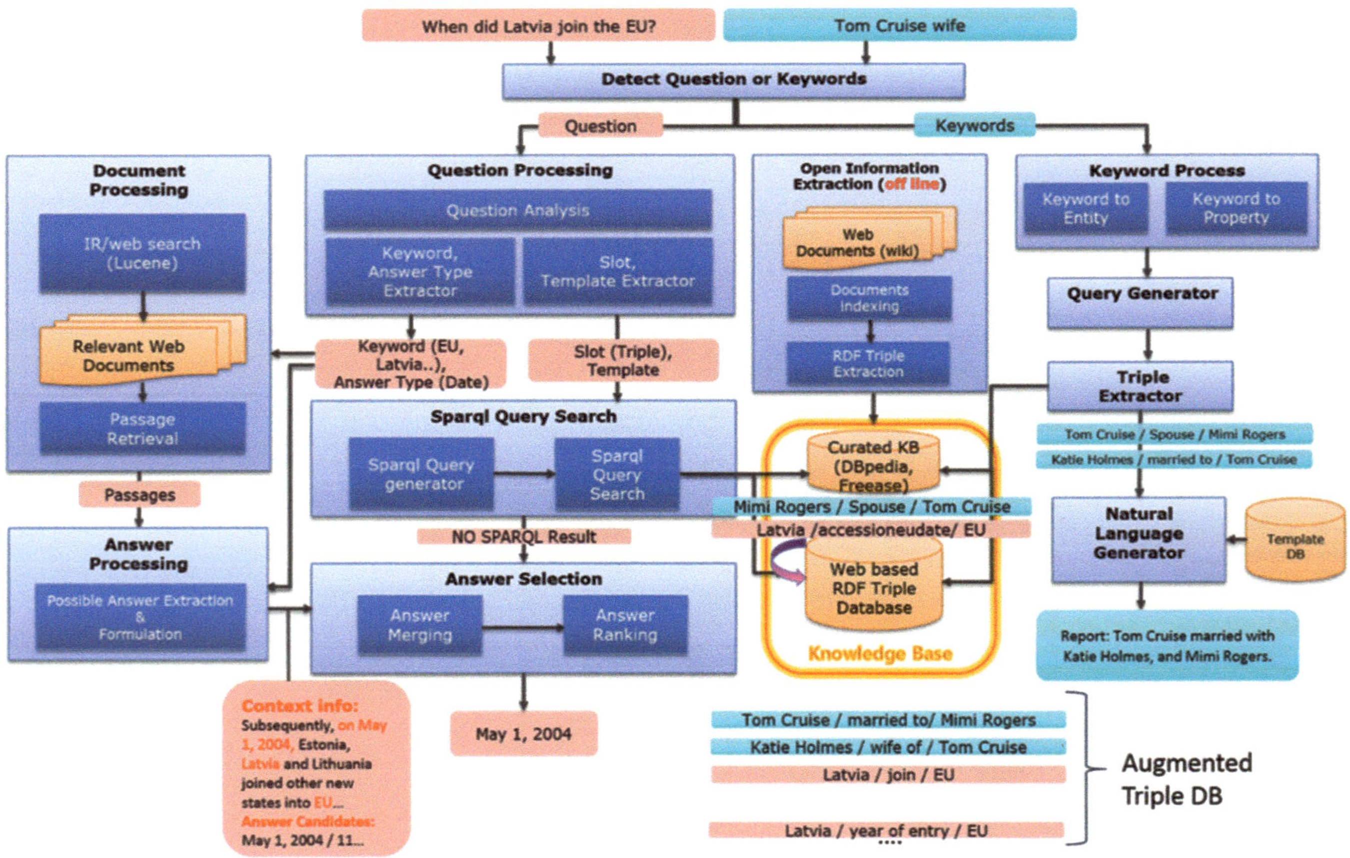

Fig. 23.1 Architecture of entire system

23.2 System Description

23.2.1 Knowledgebase Based QA

A Knowledgebase based Question Answering (KB-based QA) system that takes a natural language (NL) question as its input and retrieves its answer from structured (possibly curated) KBs like DBpedia and Freebase. A KB-based QA system uses highly structured information sources, so it produces very specific answer sets.

We take two approaches to handle the task. The first approach uses semantic parsing (Berant et al. 2013); the other uses lexico-semantic patterns (LSP) matching. In the semantic parsing approach, we first use a beam segmenter to generate candidate segmentations of an NL question; then we use string based methods and automatically generated <NL phrases, KB node mapping dictionary> to try to match KB vocabulary to the segments. We generate query candidates by using a small set of hand-crafted grammar rules to combine segments into a single formal representation of meaning. In the LSP approach, we generate patterns that consist of regular expression patterns that describe the lexical/POS/chunk type patterns of an NL question and an SPARQL query template. If a match is found, slots in the SPARQL query template are filled with the word-matched chunks from NL question. However, a KB-based QA modules has no context information and therefore cannot rank its answer candidates; instead KB-based QA passes its answer candidate to an answer merging module in the IR-based QA and lets the module rank the answer candidates.

23.2.2 Information Retrieval Based QA

An information retrieval based Question Answering (IR-based QA) system searches text to find answers. Our IR-based QA includes four modules (Fig. 23.1): the first classifies answer type and analyses the question semantically; the second retrieves passages by searching documents that are related to the user question; the third extracts answer candidates; the fourth merges answer candidates from IR-based QA and KB-based QA, scores the answer candidates and returns the final list of answers. The difference between our system and other systems is that our system uses context information to score answer candidates which are the results of the SPARQL not only from IR-based answer candidate extraction.

We used Ephyra[1] for question processing, which includes extracting keywords by lexical, syntactic and semantic analysis, and a hybrid answer-type classifier that uses rules and a classifier based on machine learning (Schlaefer et al. 2007). We also used

[1] https://www.ephyra.info.

Lucene[2] for indexing wiki pages dump and for searching documents and passages related to the answers. After passages are searched, sentences in the passages are scored. We extract named entities which have the same or similar answer types as answer candidates from n-best sentences in passages. Finally, using semantic relatedness among questions and sentences that include answer candidates, our system ranks answer candidates from IR-based and KB-based modules. At the end of that process, the system gives the final answer list to user.

23.2.3 Keyword QA

The keyword QA system is a system that takes Keyword as input and returns an NL report as the result. Because the query is a combination of keywords, multiple triples are extracted as answers. Because being required to evaluate multiple triples can confuse or irritate users, the system generates an NL report from the extracted triples.

The system matches each keyword to an entity or property in the KB to extract answer from it. To match keywords to entities or properties, we used AIDA (Hoffart et al. 2011) and ESA[3] module. First, AIDA tries to match each keyword to an entity. When keywords are not matched to an entity, we use the ESA module to match them to a property. A rule-based query generator generates a query to extract data from a KB. We used manually generated NL generation templates to generate an NL report. The NL generation templates are structured as property-predicate pairs.

We generated 670 keyword queries to evaluate the accuracy of the keyword QA module. Our system returned a correct[4] answer for 95.1 % of these questions.

23.2.4 Knowledgebase by Open Information Extraction

Although a KB has a large data capacity, it can only cover small amount of information compared to its original free text. To handle this problem, we constructed a repository that consists of triples extracted from free text. We exploit the dependency tree and semantic role labels (SRL) of a sentence to extract triples from free text.

We defined "extraction templates" that specify how triples should be extracted for each dependency tree structure pattern. To generate extraction templates automatically, we used bootstrapping methods. A whole document is retrieved to find sentences that contain word tokens that appear in arguments and relation words of

[2]http://lucene.apache.org/core.

[3]http://ticcky.github.io/esalib.

[4]"Correct" means the result was reasonable interpretation for the keyword query based on human judgment.

each seed triple. Then we constructed a dependency tree of the sentence for each seed triple, sentence pair, and identified a linear path which contains arguments and relation words. This path with position of arguments and relation words in the path can produce an extraction template.

SRL outputs similar results that can be transformed to triple format. Each predicate of the results are regarded as relation phrases and each argument and argument modifier are regarded as each argument of triples. We also used a small set of rules to transform SRL results to triples.

23.2.5 *Integration*

Our system can process both NL questions and keywords. Because NL questions and keywords are processed by different sub-systems, our system has a module that disambiguates the query form. The module identifies whether a user query is an NL question or a keyword. To disambiguate them, we trained a model based on a conditional random field algorithm. Our training data are our query data that include NL questions and keywords. Our features are an n-gram of words and POS tags.

Our system uses semantic relatedness among questions and sentences which include answer candidates to rank answer candidates from KB-based QA and IR-based QA. At the end of the process, the system gives the final answer list to user.

Acknowledgements This work was supported by ICT R&D program of MSIP/IITP [10044508, Development of Non-Symbolic Approach-based Human-Like Self-Taught Learning Intelligence Technology] and ATC (Advanced Technology Center) Program—'Development of Conversational Q&A Search Framework Based On Linked Data: Project No. 10048448'.

References

Berant J, Chou A, Frostig R, Liang P (2013) Semantic parsing on freebase from question–answer pairs. In: Empirical methods in natural language processing, October 2013, pp 1533–1544

Hoffart J, Yosef MA, Bordino I, Fürstenau H, Pinkal M, Spaniol M, Taneva B, Thater S, Weikum G (2011) Robust disambiguation of named entities in text. In: Proceedings of the conference on empirical methods in natural language processing. Association for Computational Linguistics, July 2011, pp 782–792

Schlaefer N, Ko J, Betteridge J, Pathak MA, Nyberg E, Sautter G (2007) Semantic extensions of the Ephyra QA system for TREC 2007. In: Text REtrieval conference

Chapter 24
Dialogue Platform for Interactive Personal Assistant Software

Youngmin Park, Sangwoo Kang, Myungwan Koo, and Jungyun Seo

Abstract An interactive personal assistant software system can perform services desired by users through a natural language interface. In this paper, we propose an effective knowledge platform structure that considers expanded structural application domains of language understanding and dialogue management modules. These modules form the core technology of the interactive personal assistant software. For the proposed platform, analysis factors of user intention are systematically defined to understand language, effective dialogue management methods are included to compensate for analytic errors, and the structure of ontology knowledge is described to expand domain knowledge.

Keywords Dialogue engine • Natural language processing • Dialogue platform • Dialogue manager

24.1 Introduction

Widespread use of smart mobile devices has led to an increase in the demand of a dialogue interface that comprehends user queries through natural language and provides corresponding services without requiring complicated usage procedures. However, the existing dialogue systems are used sparingly because they usually provide only low-level interaction through simple questions and answers or have a restricted structure that prevents the expansion of application domains (Harabagui et al. 2001; Lee et al. 2001).

In this paper, a dialogue platform that effectively manages dialogue and effortlessly maintains knowledge in various application domains is proposed. In Sect. 24.2, the overall structure of the dialogue platform and factors that form the platform (i.e., a language understanding model, a factor analysis model, domain knowledge, and a dialogue transition model) are explained. Finally, in Sect. 24.4, conclusions and further research plans are presented.

Y. Park • S. Kang (✉) • M. Koo • J. Seo
Department of Computer Science and Engineering, Sogang University, Seoul, South Korea
e-mail: pymnlp@gmail.com; swkang@sogang.ac.kr; mwkoo9@gmail.com; seojy@sogang.ac.kr

G. Geunbae Lee et al. (eds.), *Natural Language Dialog Systems and Intelligent Assistants*, DOI 10.1007/978-3-319-19291-8_24

24.2 Dialogue Platform for Interactive Personal Assistant Software

Figure 24.1 illustrates the platform proposed in this paper. This platform mainly consists of a dialogue engine, do engine, and knowledge manager. The dialogue engine comprises a language understanding and dialogue management model. The language understanding model analyzes user intention from their speech. The do engine manages intelligent connection services, and the knowledge manager manages knowledge of a dialogue domain. This paper mainly examines the dialogue engine and knowledge manager and excludes the do engine.

24.2.1 User Intention Analysis Factors and Language Understanding

To accurately analyze user intention through their respective speech in various application domains, the interactive personal assistant defines five analysis factors: domain, speech act, predicator, named entity, and argument.

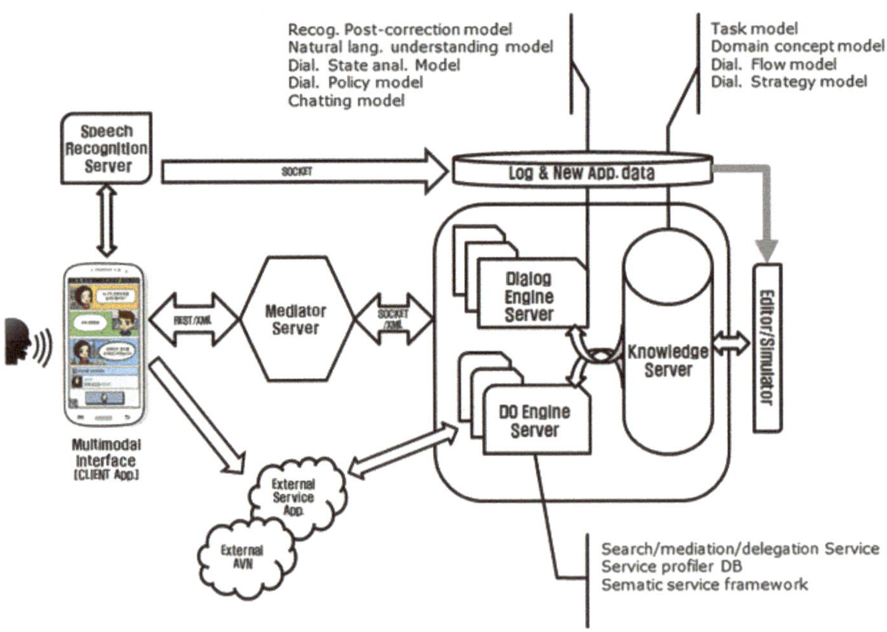

Fig. 24.1 Architecture of the platform for the personal assistant software

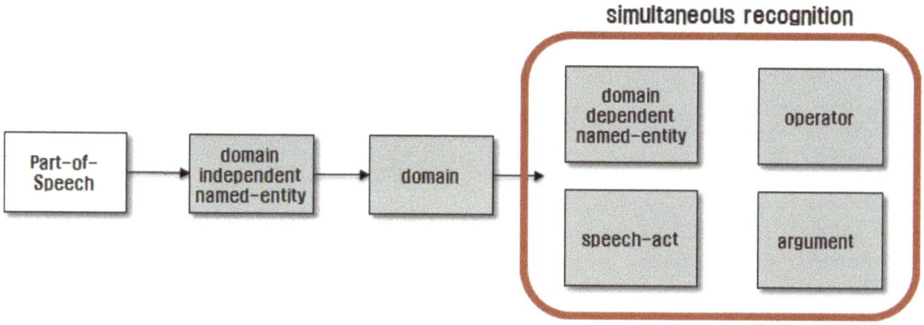

Fig. 24.2 Recognition process of language understanding

Table 24.1 State transition operations

State transition operation	Description
new_task	Beginning of a new task, creation of a state
state_changing	Factor input, adjustment of a state form
task_switching	Task switch
task_cancel	Discontinuance of dialogue, termination of a state
task_ complete	Completion of a state form
state_unknown	Failure of analysis
Exception	Exception

For language understanding, interrelated analysis factors were simultaneously analyzed to remove ambiguity. Moreover, conditional random fields, which are a part of the statistical machine learning method, were applied in the analysis model (Fig. 24.2).

24.2.2 *Interaction Model Using State Transition Operations*

The proposed platform defines actions to be performed by the system in response to user speech as state transition operations using finite-state automata-based dialogue model and determines the next operation by applying a statistical model (Table 24.1).

To estimate the state transition operations not affected by voice recognition errors, these operations are performed on each result of N-best voice recognition. Subsequently, a state transition operation (O) that statistically has the highest probability is selected. At this time, information such as results of the language understanding model (I), results of morpheme analysis (M), and dialogue status (S) is used to measure the probability, as shown in Eq. (24.1).

$$O' = \mathrm{argmax}_O P(O, I, M, S) \tag{24.1}$$

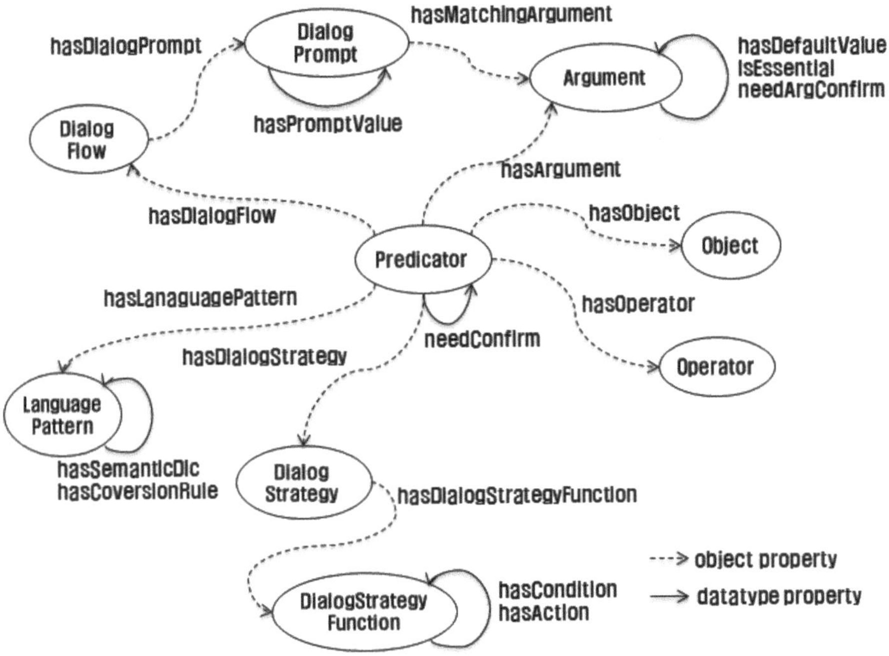

Fig. 24.3 Example of an ontology schema for domain knowledge

24.2.3 Representation of Domain Knowledge Using Ontology

A knowledge representation method based on ontology is typically used to model a lexical relation within a specific domain. It is also used to represent knowledge inside any system in a single structure (Gurevych et al. 2003).

In this study, ontology is used to represent domain knowledge that consists of domain actions and factor information required to perform each domain action. The proposed method adds a pertinent concept when a new domain or domain action is added and establishes its relation with the existing concept, thus facilitating effective domain expansion.

Figure 24.3 shows an example representing domain knowledge using ontology. The properties *has_argument* and *has_predicator* are inversely related to each other. The domain knowledge was modeled using OWL, a W3C standard.

24.3 Experiments

We evaluated usability of our model on user test. Usability is measured in the following order:

1. A user requests any one task to dialogue system, where a user requests one of the 33 designated tasks.

2. A user carries on a dialogue until the requested task finish.
3. A user estimates usability score scaled 0–5.

We evaluated for user group comprised of 20 persons. In the result of evaluation, the average usability score was 4.42 and the average number of turns was 4.36 (min = 2, max = 14).

24.4 Conclusion

In this paper, a dialogue platform for interactive personal assistant software was proposed. The proposed model can effectively expand domain knowledge using ontology and process dialogues without being strongly affected by voice recognition errors.

Acknowledgments This work was supported by the Basic Science Research Program through the National Research Foundation of Korea (NRF) funded by the Ministry of Education (No. NRF-2013R1A1A2010190).

References

Gurevych I, Porzel R, Slinko E, Pfleger N, Alexandersson J, Merten S (2003) Less is more: using a single knowledge representation in dialogue systems. In: Proceedings of the HLT-NAACL'03 workshop on software engineering and architecture of language technology systems (SEALTS)

Harabagui S, Moldovan D, Pasca M, Rada M, Surdeanu M, Bunescu R, Girju R, Rus V, Mororescu P (2001) The role of Lexico-Semantic feedback in open-domain textual question-answering. In: Proceedings of the 39th annual meeting of the association for computational linguistics (ACL-2001), Toulouse, France, pp 274–281

Lee GG, Seo J, Lee S, Jung H, Cho B, Lee C, Kwak B, Kim H, Kim K (2001) SiteQ: engineering high performance QA system using Lexico-Semantic pattern matching and shallow NLP. In: Proceedings of the 10th text retrieval conference (TREC-10)

Chapter 25
Performance Analysis of FFNN-Based Language Model in Contrast with *n*-Gram

Kwang-Ho Kim, Donghyun Lee, Minkyu Lim, Minho Ryang, Gil-Jin Jang, Jeong-Sik Park, and Ji-Hwan Kim

Abstract In this paper, we analyze the performance of feed forward neural network (FFNN)-based language model in contrast with *n*-gram. The probability of *n*-gram language model was estimated based on the statistics of word sequences. The FFNN-based language model was structured by three hidden layers, 500 hidden units per each hidden layer, and 30 dimension word embedding. The performance of FFNN-based language model is better than that of *n*-gram by 1.5 % in terms of WER on the English WSJ domain.

Keywords Feed forward neural network • *n*-Gram • Language model • Performance analysis

25.1 Introduction

The language model is constructed based on the relationship between words. The representative language model, called *n*-gram language model, has the following problems: (1) The probability related to unseen data is estimated by the smoothing techniques such as back-off and interpolation, but these techniques are unreliable. (2) This method can only express information based on a restricted word history,

K.-H. Kim • D. Lee • M. Lim • M. Ryang • J.-H. Kim (✉)
Department of Computer Science and Engineering, Sogang University,
Seoul 121-742, South Korea
e-mail: kimkwangho@sogang.ac.kr; redizard@sogang.ac.kr; lmkhi@sogang.ac.kr; k2minho@sogang.ac.kr; kimjihwan@sogang.ac.kr

G.-J. Jang
School of Electronics Engineering, Kyungpook National University,
Daegu 702-701, South Korea
e-mail: gjang@knu.ac.kr

J.-S. Park
Department of Information and Communication Engineering,
Yeungnam University, Gyeongsan 712-749, South Korea
e-mail: parkjs@yu.ac.kr

G. Geunbae Lee et al. (eds.), *Natural Language Dialog Systems and Intelligent Assistants*, DOI 10.1007/978-3-319-19291-8_25

because of limitations in n. For a fixed amount of training data, the amount of unseen data increases as the size of n also increases. In spite of these problems, n-gram is used widely in the speech recognition field as the representative language model. The trigram or 4-gram models in particular are most commonly used in speech recognition. For solving n-gram based language model problems, many researches focus on feed forward neural network (FFNN)-based language models.

In this paper, we investigate the performance analysis of FFNN-based language model in contrast with n-gram. This paper is organized as follows. Section 25.2 introduces the related works about n-gram-based language model. Section 25.3 describes FFNN-based language model. Section 25.4 concludes this paper.

25.2 Related Works

A language model makes a guess of the current word based on the previous word sequence. For the word sequence $W = (w_1, w_2, \ldots, w_N)$, the probability of a language model is denoted as $p(W)$.

$$p(W) = p(w_1, w_2, \ldots, w_N) = \prod_{i=1}^{N} p\left(w_i \middle| w_1, \ldots, w_{i-1}\right) \tag{25.1}$$

When the number of words in the word history $(w_1, \ldots, w_{i-1})$ increases, it becomes increasingly difficult to calculate the probability for the current word w_i because the word history may not appear in the text corpus. For this reason, Markov assumption is applied to the language model to compute $p(w_i|w_1, \ldots, w_{i-1})$. Here, the number of words influencing the current word w_i, is $n - 1$. This can be expressed in the n-gram, which is widely used on language modeling. Equation (25.2) is a restatement of Eq. (25.1) after Markov assumption is applied to the language model.

$$\prod_{i=1}^{N} p\left(w_i \middle| w_1, \ldots, w_{i-1}\right) \approx \prod_{i=1}^{N} p\left(w_i \middle| w_{i-n+1}, \ldots, w_{i-1}\right) \tag{25.2}$$

Take a trigram example to calculate the probabilities for the input "be or not". The probability for this trigram can be expressed as p ("not"|"be", "or"). To estimate the probability of the trigram, the frequency of the word sequence consisting of three consecutive words from the text corpus is calculated. The calculated frequency is normalized to the total sum of the frequencies of the trigrams containing the shared word history of ("be", "or"). A specific n-gram probability can be expressed as Eq. (25.3):

$$p\left(w_i \middle| w_{i-n+1}, \ldots, w_{i-1}\right) = \frac{C\left(w_{i-n+1}, \ldots, w_{i-1}, w_i\right)}{\sum_{k=1}^{|V|} C\left(w_{i-n+1}, \ldots, w_{i-1}, w_i\right)} \tag{25.3}$$

The word sequence frequency of "$w_{i-n+1}, \ldots, w_{i-1}, w_i$" from the training text corpus is denoted by $C(w_{i-n+1}, w_{i-1}, w_i)$. V is the set of recognizable words while $|V|$ is the total number of recognizable words. Statistical word-based n-gram language model is not trained with the entire text corpus which contains all of the possible word sequences. Since all possible word sequences are not used in training, the problem of unseen word sequence inevitably arises. One solution, called smoothing, assigns nonzero low probability value to word sequence frequencies which originally had zero probability. Even with this smoothing method, stable probability estimation for unseen word sequence cannot be completed. This reveals the limitation of the n-gram language model.

25.3 FFNN-Based Language Model

FFNN structure using multilayer perceptron (Bishop 1995) was proposed in Bengio et al. (2003) and Bengio (2009), with which a language model was generated to be applied to speech recognition. Language model learning in the telephone conversation domain and its performance evaluation were presented in Schwenk and Gauvain (2002, 2004). The input and output of FFNN in the FFNN structure are as follows: For input, a word history consists of $n-1$ words in n-gram, and each word in the word history is represented in the 1-of-$|V|$ format. Therefore, it can be said that the set of all of the words in set V is collectively expressed as a vector where the number of its constituents is $|V|$. Here, the index of the word is assigned as "1" and the other values with "0". To reduce the vector dimension of word vector, a projection matrix is applied. A projection matrix expresses each word in vector regardless of the word's temporal location in the word history. For output, the probability of the current word is estimated by the given word history; the current possible words consist of the recognizable words in V. As a result, output size is expressed as $|V|$.

The methods using short list, regrouping, and block model in Schwenk and Gauvain (2005) reduced the training time for FFNN-based language model for large corpus. In shortlist method, only the most frequent word subset in the training data is put in the output layer word list so as to reduce the size of the output layer. In regrouping, while the probability of the current word is being calculated in the word history, the current words are grouped together so that a sequence of words is trained as one word. In block mode, some vectors were merged into one matrix, called a vector block. These three methods were applied to reduce the training time. The efficacy and performance of these methods were evaluated with 65K words in a French broadcast news. The performance of FFNN-based language model is better than that of n-gram by 0.63 % in terms of WER on the French broadcast news. The number of output layers was reduced with shortlist method, but the improvement in performance was insignificant and the reduction in training time was investigated clearly.

In Arisoy et al. (2012), the performance of language model with varying number of hidden layers (one to four) on FFNN language model was measured. In the experiment, performance variances were monitored with different numbers of hidden layers, hidden units. The FFNN-based language model was structured by three hidden layers, 500 hidden units per each hidden layer, and 30 word embedding each word. The performance of FFNN-based language model is better than that of n-gram by 1.5 % in terms of WER on the English WSJ domain. The training corpus consists of 23.5 million words from 900K sentences.

25.4 Conclusions

In this paper, we analyzed the performance of FFNN-based language model in contrast with n-gram. The probability of n-gram language model was estimated based on the statistics of word sequences. The FFNN-based language model was structured by three hidden layers, 500 hidden units per each hidden layer, and 30 dimension word embedding. The performance of FFNN-based language model is better than that of n-gram by 1.5 % in terms of WER on the English WSJ domain. The training corpus consists of 23.5 million words from 900K sentences.

Acknowledgments This research was supported by Basic Science Research Program through the National Research Foundation of Korea (NRF) funded by the Ministry of Science, ICT and Future Planning (No. NRF-2014R1A1A1002197).

References

Arisoy E, Sainath T, Kingsbury B, Ramabhadran B (2012) Deep neural network language models. In: NAACL-HLT 2012 workshop, pp 20–28

Bengio Y (2009) Learning deep architectures for AI. J Found Trends Mach Learn 2:1–127

Bengio Y, Ducharme R, Vincent P, Jauvin C (2003) A neural probabilistic language model. J Mach Learn Res 3:1137–1155

Bishop C (1995) Neural networks for pattern recognition. Clarendon, Oxford

Schwenk H, Gauvain J (2002) Connectionist language modeling for large vocabulary continuous speech recognition. In: International conference on acoustics, speech and signal processing, pp 765–768

Schwenk H, Gauvain J (2004) Neural network language models for conversational speech recognition. In: International conference on speech and language processing, pp 1215–1218

Schwenk H, Gauvain J (2005) Training neural network language models on very large corpora. In: Empirical methods in natural language processing, pp 201–208

Chapter 26
GenieTutor: A Computer-Assisted Second-Language Learning System Based on Spoken Language Understanding

Oh-Woog Kwon, Kiyoung Lee, Yoon-Hyung Roh, Jin-Xia Huang, Sung-Kwon Choi, Young-Kil Kim, Hyung Bae Jeon, Yoo Rhee Oh, Yun-Kyung Lee, Byung Ok Kang, Euisok Chung, Jeon Gue Park, and Yunkeun Lee

Abstract This paper introduces a computer-assisted second-language learning system using spoken language understanding. The system consists of automatic speech recognition, semantic/grammar correction evaluation, and tutoring module. The speech recognition is optimized for non-natives as well as natives for educational purpose and smooth interaction. Semantic/grammar correction evaluation evaluates whether the non-native learner's utterance is appropriate semantically and is correct grammatically. Tutoring module decides to go to the next turn or ask the learner to try again, and also provides a turn-by-turn corrective feedback using evaluation results. We constructed English learning service consisting of three stages such as Pronunciation Clinic, Think&Talk and Look&Talk using the system.

Keywords Computer-assisted second-language learning system • Non-native-optimized speech recognition • Grammar error correction • Semantic correctness evaluation • Educational feedback

26.1 Overview

This demonstration paper introduces GenieTutor—a computer-assisted second-language learning system using spoken dialog processing technology. GenieTutor plays the role of a language (English at present) tutor by automatically correcting grammar and checking content properness of the learners' responses and giving

O.-W. Kwon (✉) • K. Lee • Y.-H. Roh • J.-X. Huang • S.-K. Choi • Y.-K. Kim • H.B. Jeon
Y.R. Oh • Y.-K. Lee • B.O. Kang • E. Chung • J.G. Park • Y. Lee
Automatic Speech Translation and Artificial Intelligence Research Center, Electronics and Telecommunications Research Institute, 218 Gajeong-ro, Yuseong-gu, Deajeon 305-700, South Korea
e-mail: ohwoog@etri.re.kr

G. Geunbae Lee et al. (eds.), *Natural Language Dialog Systems and Intelligent Assistants*, DOI 10.1007/978-3-319-19291-8_26

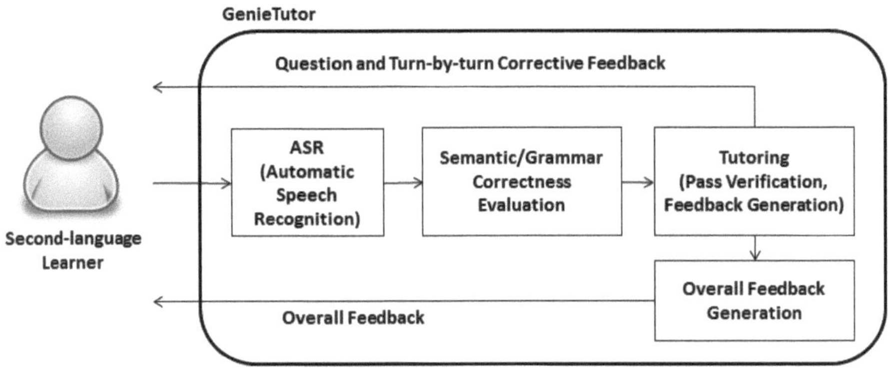

Fig. 26.1 Schematic diagram of GenieTutor

educational feedbacks to learners. The speech recognition system is optimized for non-natives as well as natives for educational purpose and smooth interaction.

GenieTutor leads dialogs with learners to focus on the certain topics by asking questions and providing suggestions for the answers. The system recognizes the speech which is answered in second-language, evaluates if it is the proper answer for given question, checks grammatical errors, and provides feedbacks to help learners practice their English proficiency. Figure 26.1 shows the schematic diagram of GenieTutor.

26.1.1 Non-Native-Optimized Speech Recognition

In order to construct non-native-optimized speech recognition system, we used native English utterances (380 h in total) and non-native Korean spoken English utterances (408 h in total) where each utterance was sampled at a rate of 16 kHz. We first extracted speech feature vectors at every 10 ms for a 20 ms analysis window (Chung et al. 2014; Lee et al. 2014). Then, we trained separately native acoustic models (AMs) and non-native AMs by using native and non-native utterances, respectively (Young et al. 2009). The AMs are composed of 3-state, 16-mixture, and cross-word triphone models. Next, native and non-native AMs were merged based on Gaussian mixture models based on multi-space probability distribution.

In addition, we collected about 130 million English text sentences including the English scripts of GenieTutor, the spontaneous sentences, and the grammatically wrong sentences that are commonly occurred by Korean learners. Then, we constructed a back-off trigram language model with 54,826 most frequent words.

26.1.2 Evaluation and Tutoring

Evaluation module evaluates whether the learner's utterance is appropriate semantically and is correct grammatically. The semantic correctness checker decides whether to pass the learner on the current turn or not, using the domain knowledge and language model. The semantic correctness is classified into six categories such as "perfect", "too few modifiers", "inflection error", "subject-verb error", "content error", and "illegal expression", for the feedback to the learner in the tutoring step.

Grammatical error correction plays an important role in second language learning using computers. Many grammatical error correction systems aim at detecting and correcting grammar errors in essays written by students who are non-native speakers of English (Ng et al. 2014). We focus on grammar errors in dialogue between a student and a learning system. For a grammatical evaluation, we employ a grammatical error correction system which is composed of three different approaches: a rule-based, a machine learning based, and an *n*-gram based correction modules. According to the performance of each correction module for grammatical error types, we assigned a proper weight to correction candidates by each correction module. To suppress false alarms which are critical to secondary language tutor system, the correction system has a voting strategy to filtering implausible correction candidates.

Tutoring module decides to go to the next turn or ask the learner to try again according to the semantic correctness evaluation and also provides a turn-by-turn corrective feedback using evaluation results. The feedback consists of three parts that are shown to the learner in a step-by-step and sequential manner. The first part briefly shows a grade of pass or fail and the words with grammar errors. The second part suggests some recommendation sentences with reasons of failure when the semantics of the input utterance is not appropriate to the current question. The last part is the corrective feedback of grammatical errors described in the first part.

Once the dialogues between the GenieTutor and the learner finished, overall evaluation module assesses the English learner's performance and produces feedbacks with the scores to show which part the learner should focus more on. Several measurements are adopted for the evaluation, including task proficiency, grammar accuracy, vocabulary diversity, and grammar complexity. Task proficiency evaluates how fluently the conversation has been maintained according to the numbers of pass and failing turns. Grammar accuracy is scored based on the number of grammar errors, which is checked by grammar check module. Grammar complexity compares user's sentences with the references offered by native speakers and gives relative complexity score by considering the length and the number of conjunctions. For the vocabulary diversity the system is supposed to check if the learners tend to use the same expressions or words from elementary level vocabularies when there are better alternatives and provides synonyms and similar expressions to improve learners' vocabulary.

26.2 Curriculum of GenieTutor

The service or curriculum of GenieTutor consists of three stages such as Pronunciation Clinic, Think&Talk, and Look&Talk.

26.2.1 Pronunciation Clinic

The Pronunciation Clinic provides the English pronunciation skills that are specialized for Korean non-native language learners and the corresponding speaking practices based on speech recognition. As shown in Fig. 26.2, the Pronunciation Clinic consists of four steps: (a) selecting a lesson among the 30 Pronunciation Clinic lessons, (b) listening a video lecture for the selected lesson, (c) selecting the words or sentences that contain the target pronunciations of the selected lesson, and (d) practicing a spoken dialogue by using the selected words or sentences.

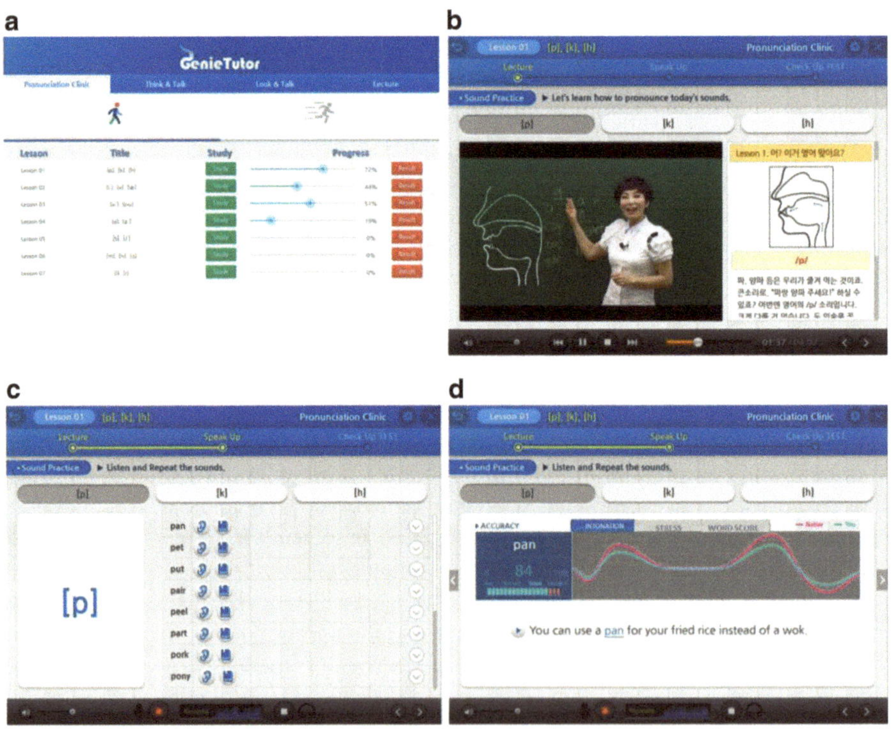

Fig. 26.2 Four steps of Pronunciation Clinic

Especially, the last speaking practice step provides pronunciation scores and the comparative analysis results of the intonation and stresses between the learners' and the corresponding native utterance.

26.2.2 Think&Talk: Talk with the Computer on Various Subjects

In the stage Think&Talk an English learner is supposed to select a dialog context from a pool of domains and its contents and to practice the dialog utterance that corresponds to the context. If the learner makes a semantic or grammatical mistake, GenieTutor gives him some corresponding feedbacks and asks him to try again a correct utterance (Fig. 26.3).

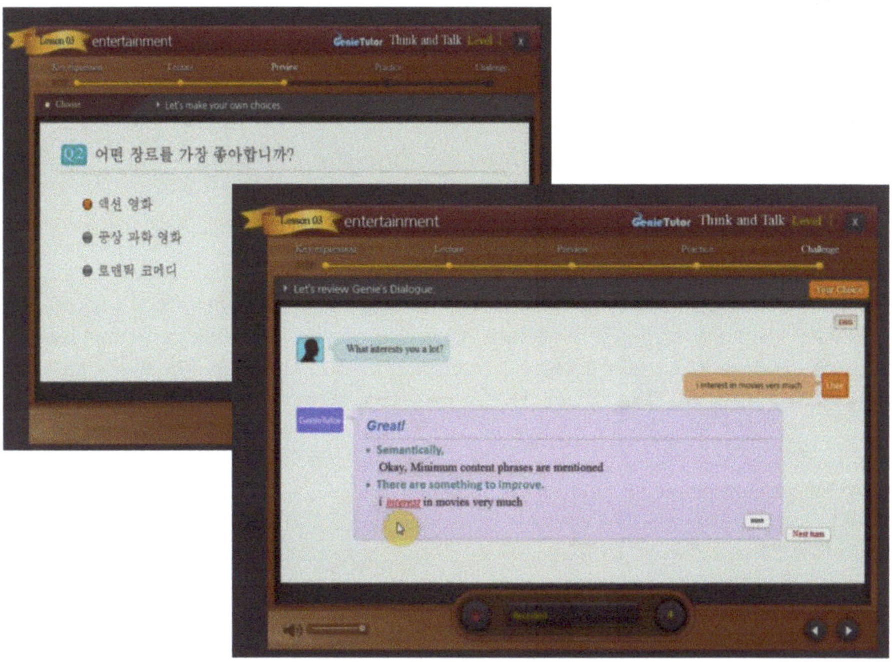

Fig. 26.3 Instance selection and a dialogue exercise in the stage Think&Talk

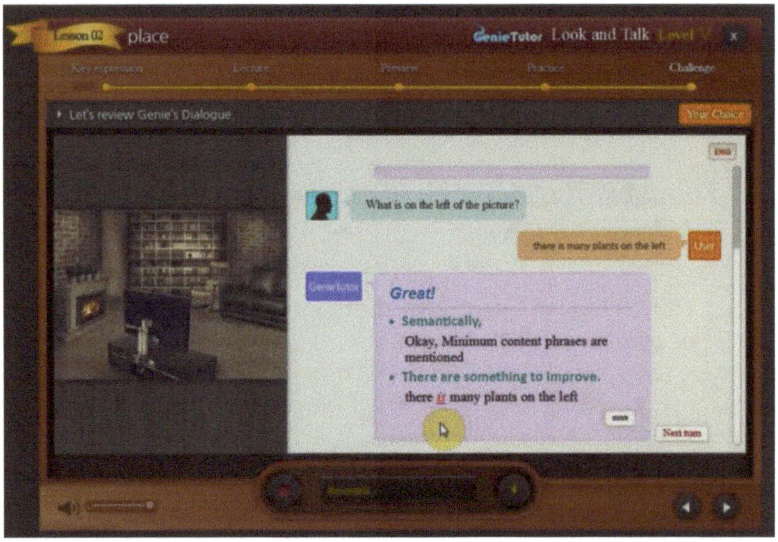

Fig. 26.4 A dialogue exercise of the stage Look&Talk

26.2.3 Look&Talk: Look and Describe the Pictures to the Computer

In the stage Look&Talk an English learner is supposed to watch and describe the picture according to the system's question to enhance his/her power of expression. Likewise with the stage Think&Talk, GenieTutor gives feedbacks to the learner and asks him to try again a correct utterance if the learner makes a semantic or grammatical mistake (Fig. 26.4).

Acknowledgements This work was supported by the ICT R&D program of MSIP/IITP (10035252, Development of dialog-based spontaneous speech interface technology on mobile platform).

References

Chung H, Lee SJ, Lee YK (2014) Weighted finite state transducer-based endpoint detection using probabilistic decision logic. ETRI J 36:714–720

Lee SJ, Kang BO, Chung H, Lee YK (2014) Intra- and inter-frame features for automatic speech recognition. ETRI J 36:514–517

Ng HT, Wu SM, Briscoe T (2014) The CoNLL-2014 shared task on grammatical error correction. In: Proceedings of the 18th conference on computational natural language learning, pp 1–14

Young S et al. (2009) The HTK book (for HTK version 3.4). Microsoft Corporation, Cambridge University Engineering Department, March 2009

Chapter 27
Learning Speed Improvement Using Multi-GPUs on DNN-Based Acoustic Model Training in Korean Intelligent Personal Assistant

Donghyun Lee, Kwang-Ho Kim, Hee-Eun Kang, Sang-Ho Wang, Sung-Yong Park, and Ji-Hwan Kim

Abstract This paper proposes a learning speed improvement using multi-GPUs on DNN-based acoustic model training in Korean intelligent personal assistant (IPA). DNN learning involves iterative, stochastic parameter updates. These updates depend on the previous updates. The proposed method provides a distributed computing for DNN learning. DNN-based acoustic models are trained by using 320 h length Korean speech corpus. It was shown that the learning speed becomes five times faster on this implementation while maintaining speech recognition rate.

Keywords Deep neural network • Graphical processing unit • Amazon elastic compute cloud • Acoustic model

27.1 Introduction

Speech interface has been performing an important role in enabling human–computer interactions. For example, intelligent personal assistants (IPAs) such as Siri in iOS system and Google Now in the android system provide a hands free, eyes free solution to search the web and reserve restaurants using speech interface (Deng et al. 2002). This interface is becoming steadily popular and is being used to simplify the interaction between users and mobile devices due to its improved performance of automatic speech recognition (ASR) (Wang 2014). This improved performance is mainly caused by using automatic retraining of models which are used in ASR.

D. Lee • K.-H. Kim • H.-E. Kang • S.-H. Wang • S.-Y. Park • J.-H. Kim (✉)
Department of Computer Science and Engineering, Sogang University, Seoul 121-742, South Korea
e-mail: redizard@sogang.ac.kr; kimkwangho@sogang.ac.kr; heun831@sogang.ac.kr; wangsh@sogang.ac.kr; parksy@sogang.ac.kr; kimjihwan@sogang.ac.kr

G. Geunbae Lee et al. (eds.), *Natural Language Dialog Systems and Intelligent Assistants*, DOI 10.1007/978-3-319-19291-8_27

The goal of an ASR system is to convert speech information into text symbols of the spoken words. The system computes a series of most plausible words *W*, from an acoustic vector series *X*. Since the number of probable *X* is infinite, the following Bayes theorem is used as in (27.1):

$$\widehat{W} = argmax_w P\left(W \middle| X\right) = argmax_w \frac{P(W)P\left(X \middle| W\right)}{P(X)} \approx argmax_w P(W)P\left(X \middle| W\right) \tag{27.1}$$

In (27.1), the probability of acoustic vector series *X* is the denominator term *P*(*X*). It can be disregarded since it does not depend on the *W*. The probability for *W* is *P*(*W*). It is calculated from the language model (LM). The LM assigns a probability to a series of word symbols uses the LM. The LM serves as a guide for predicting the next word when word history is given. $P(X|W)$ is a conditional probability which is to be calculated by an acoustic model (AM) for *W*. The AM is to model speech units such as words or syllables by using their acoustic properties. Also, before constructing the AM and LM of the ASR system, it is necessary to choose a group of words upon the domain of corpus. This group of words is named the lexicon of the ASR system. Then, the decoder of the ASR system searches a word series $\hat{W}$, which is the result of the highest product of *P*(*W*) and $P(X|W)$.

After the development of the ASR system, practically, hidden Markov models (HMMs) are still used in the generation of AMs in the most of ASR systems. An HMM is a statistical model for producing a sequence of symbols. There are widely used two of many ASR systems depending on how probabilities of HMM are modeled. One is the Gaussian mixture model/hidden Markov model (GMM/HMM) system and the other is the deep neural network/hidden Markov model (DNN/HMM) system (Bourlard and Morgan 1994).

Basically, the DNN/HMM system has some advantages over the GMM/HMM system (Bourlard and Morgan 1994): (1) DNNs are naturally more suitable for a discriminative training than GMMs. The discriminative training is a method for training models to minimize the error rate while maximizing the distance between the correct model and its another model. (2) DNNs can include multiple constraints. In other words, unlike GMM/HMM systems, there is no need for strict assumptions about statistical distributions of features used in DNN/HMM systems. (3) DNNs are typically higher parallel and regular structures than GMMs. These structures make DNNs amenable to high-performance architectures. However, these apparent advantages of using DNN/HMM systems for modeling AMs are required to spend much time in DNN learning.

To improve speed of DNN learning, some researchers have used multi-core CPU system instead of single-core CPU system to make DNN learning as parallel task learning (Seltzer and Droppo 2013; Tur 2006; Li et al. 2011). However, the number of CPU cores was limited, because the price of CPU cores and power consumption had high cost. On the other hand, the hardware of GPU can work simultaneously with thousands of threads on the available cores with a few overhead. It makes GPU very suitable for parallel computing especially in iterative and simple computation with small data. Since DNN learning inherently has iterative, stochastic parameter updates, GPUs have received attention to use its many cores for parallel computing with small data (Oh and Jung 2004; Steinkraus et al. 2005; Raina et al. 2009).

However, to use a GPU effectively, the size of the parameters or the data should be reduced to avoid CPU-to-GPU bottleneck. The main bottleneck is generated in memory transferring between RAM and GPU's global memory. According to an experiment, multiplying two 1000×1000 matrices using their GPU configuration took 20 ms, but the actual computation occupied only 0.5 % of that time, and the rest of time was used for transferring data in and out from global memory to RAM (Raina et al. 2009).

Therefore, we focus on distributed computing methods such as MapReduce and grid engine (GE) to minimize CPU–GPU bottleneck. We expect to speed improvement of DNN learning by distributing several computations and data to several clusters. If memory transfers are performed only in large batch, bottleneck between RAM and GPU will be reduced. MapReduce is one of the data-parallel schemes and it is not an appropriate framework for DNN learning, because DNN learning involves iterative, stochastic parameter updates, where any update depends on the previous updates. For this reason, we decide to use GE, a parallel tasking platform well suited for the various computations including iterative computations, to speed up DNN learning by using distributed processing.

This paper proposes a multi-GPU architecture using GE to improve training speed of AM trained by the DNN/HMM system. The proposed architecture is based on GE which is a parallel tasking platform well suited for the various computations including iterative computations. From this reason, proposed architecture solves the problem in using MapReduce for DNN learning.

The rest of this paper is organized as follows. Section 27.2 introduces the related work about improvement of machine learning algorithm. Section 27.3 describes a multi-GPU architecture using GE and illustrates an experiment to estimate learning speed of the Korean acoustic model. Section 27.4 concludes this paper.

27.2 Related Work

DNNs have been shown to outperform GMMs on various applications using the speech recognition technology. Since around 2010, many researches have been performed in speech recognition fields, and some of the companies such as Google and Microsoft have been starting to apply DNNs into their speech recognition applications.

Previous studies regarding DNNs learning have been conducted by a single server. A DNN-based model gives good results, but its learning speed is very slow. For speed improvement of DNN learning, many studies have shown that parallel task learning has been used (Seltzer and Droppo 2013; Tur 2006; Li et al. 2011). To enable this, multi-core systems have been used, but there has been limitation on the number of cores.

For this reason, many researchers got interested in using GPU to use many cores. Thousands of threads in GPU can be performed concurrently and scheduled where cores are available with little overhead. Because of such parallelism, GPUs attract attention as GPGPU. Especially, GPUs can concurrently perform the matrix operations like matrix multiplication, and such parallelization scheme can compute sigmoid function in DNNs. DBN learning speed using GPU is 72.6× faster than using dual-core CPU (Raina et al. 2009). Moreover, Kyong and Jung improve the time performance of a text detection system by implementing the matrix multiplication of a neural network (Kyong and Jung 2004). They used the parallelism of a GPU by amassing many input feature vectors and weight vectors, then transforming a lot of the inner-product operations into one matrix operation. Steinkraus et al. (2005) implement a generic 2-layer fully connected neural network by using GPU and achieve 3× speedup for both training and testing. Many authors enhanced time performance of training and learning by using GPUs. However, to use a GPU effectively, we had to reduce the size of the parameters or the data to avoid CPU-to-GPU bottleneck. Therefore, distributed computing methods such as MapReduce have begun to receive attention.

In "data-parallel" schemes perspective, DNNs' standard algorithm is difficult to be parallelized because DNNs involve iterative computations and stochastic parameter updates. Each update has a dependency on the previous. If data-parallel schemes are used in standard DNNs, information is inevitably lost when concurrently computing the updates. MapReduce is one of the data-parallel schemes, so it is not an appropriate framework for DNN learning.

Recently, for DNN training speed improvement, grid computing was focused by many researchers. Grid computing offers a configurable environment and can potentially provide applications with an architecture for easy and transparent access to geographically distributed heterogeneous resources, like data storage, networks, and computational resources across different organizations and administrative domains (Reed et al. 2003; Foster and Kesselan 1999; Allcock et al. 2002). The grid computing provides DNN training methods: (1) a DNN training method with different initial training parameters on each node and (2) a DNN training method with training sets, parts of split total training set, on each node (Torresen and Landsverk 1998; Castellano et al. 2009). The method (2) showed better performances in terms of accuracy and training speed than performances of the method (1) (Castellano et al. 2009).

A GE is one of the platforms in the grid computing and it provides the above method (2). For this reason, we determine to adopt GE to speed up DNN learning by using distributed processing.

Also, GE enables each cluster to use GPU. We select sun grid engine (SGE) provided by Sun Microsystems, Inc. since it is serviced as open source software. SGE is a resource management software to accept jobs offered by users and schedule those jobs for execution on appropriate systems in the grid based upon resource management policies (Dragan 2002). Users are able to submit literally millions of jobs at a time without being anxious about where they run.

27.3 Experiment

For DNN-based acoustic model training data, approximately 160 K utterances were used, which were recorded at 16-bit resolution with a sampling rate of 16 kHz in one channel.

The speech recognition performance is measured with recognition accuracy as in Eq. (27.2), where N, H, and I denote the number of words in the text, correctly recognized hit words, and incorrectly inserted words.

$$\text{Accuracy} = (H - I)\,/N \times 100\ \% \tag{27.2}$$

The learning speed improvement on DNN-based acoustic model is measured by total learning time for DNN-based acoustic model learning.

Section 27.3.1 describes the SGE architecture which is used in the proposed method. Section 27.3.2 shows experimental results.

27.3.1 SGE Architecture in the Proposed Method

This section describes a distributed architecture using open source software, SGE. SGE is a batch scheduler software which distributes jobs submitted by users to the cluster nodes. Main components of SGE are as follows: a master host, submit hosts, and execution hosts (slave). The overall architecture of SGE is shown in Fig. 27.1.

Users submit jobs to the queue via the submit host. The master host selects one of the execution hosts which has the least load and assigns jobs to it.

The number of jobs that can be concurrently processed by SGE is equal to the sum of the execution hosts' cores. If the number of jobs submitted to the job queue is greater than the total number of the execution hosts' cores, the remaining jobs are waiting in the queue and will be assigned to the execution host if available. According to the default setting of SGE, the master host is used as a submit host at the same time.

For DNN learning, a master host is split into a number of execution hosts from the server that stores the entire learning data. Subsequently, jobs are allocated to all execution hosts. In this case, a job corresponds to a command that specifies which learning data each execution host must copy out of the split learning data. In the currently implemented DNN learning system, when the master host splits the whole learning data, an index is assigned sequentially to the split data. Furthermore, when the execution hosts copy the split learning data, the learning data that has the same index as that assigned to the execution host is copied.

When all execution hosts have finished copying data, the master host sends a job to each execution host to perform DNN learning by using the copied learning data. All execution hosts perform DNN learning through the job sent from the master host. In this case, all execution hosts perform DNN learning using the same DNN structure. For example, assuming all split learning data are mutually independent, if 10 execution hosts perform the DNN learning, 10 DNN models, which have mutually different parameter values, are created.

While the execution hosts are carrying out the DNN learning, the master host sends a job to each execution host to send the updated parameter information to the master host once the DNN learning is completed. At the completion of DNN learning, the execution hosts send the updated DNN parameter information to the master host immediately, without waiting for the other execution hosts to finish the DNN learning.

In the currently implemented DNN learning system, an index is assigned to the data which is split by the master host with the whole learning data. In the implemented system, updated DNN parameter information is received from 10 execution hosts, and after calculating their sum, the average value is calculated by dividing the sum by 10.

The DNN structure in the implemented system consists of one input layer, four hidden layers, and one output layer. The number of nodes is 440 for the input layer, 400 for the hidden layer, and 6260 for the output layer. In this case, the number of parameters to be learned is about 3 million.

The total learning data used for the test in this study are 320 h, and assuming that 100 samples are viewed per second, the required total number of calculations is 382T.

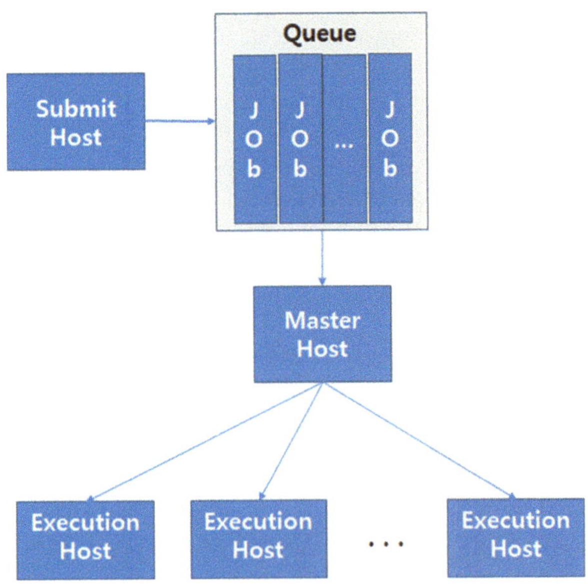

Fig. 27.1 Overview architecture of SGE

Table 27.1 Detailed specifications of Amazon EC2 servers

Role	Detailed specification	Amazon EC2 server type
Master	Number of cores: 32Memory capacity: 244 GBNo. of GPU: 0	r3.8xlarge
Slave	Number of cores: 8Memory capacity: 15 GBNo. of GPU: 1	g2.2xlarge

Table 27.2 Comparison of the acoustic model performance and learning speed by server architectures

Server architecture for learning	WER (man, %)	WER (woman, %)	Learning speed (h)
Single server (one slave)	41.34	43.26	40
Multiservers (5 slaves using GE)	41.25	43.22	8

This architecture is implemented by using Amazon Elastic Compute Cloud (Amazon EC2). The Master server utilizes 32 CPU cores and 244 GB memory, and each of the Slave servers uses 8 CPU cores, 15 GB memory, and one GPU. A detail of regarding specifications of the servers is shown in Table 27.1.

27.3.2 Experimental Result

For evaluating DNN-based acoustic model, 1.7 K utterances were used for Korean speech evaluation corpus. 860 utterances were spoken by one man and 840 utterances were spoken by one woman. These corpus were recorded with a sampling rate of 16 kHz in one channel at 16-bit resolution.

We experimented on a distributed learning about the DNN-based Korean acoustic model using multiple GPUs on Amazon EC2. For a comparative experiment, we conducted a same experiment by using a single server which has the same specification with the slave server of Table 27.1. A result of the experiment is given in Table 27.2.

There is few difference in WER between the two architectures. For the learning speed, however, DNN-based Korean acoustic model learning using 5 slaves with 5 GPUs has shown about 5 times faster than the learning speed of one slave with one GPU. This result means that it is possible to improve the learning speed while performance of the acoustic model is maintained if an acoustic model is learned by multiple GPUs based on GE.

Figure 27.2 summarizes the DNN-based acoustic model learning speed according to number of servers. In Fig. 27.2, multiservers with 10 slaves (10 GPUs) have shown about 10 times faster than the learning speed of one slave with one GPU.

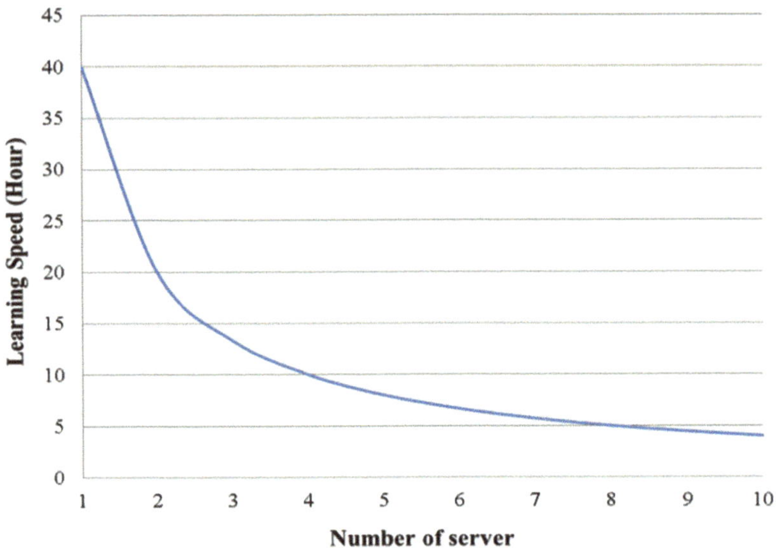

Fig. 27.2 DNN-based acoustic model learning speed according to the number of servers

The proposed method using multi-GPUs provides superior learning speed improvement to the single server using one GPU. The proposed method produces a level of accuracy similar to single server using one GPU. Therefore, it is concluded that the proposed multi-GPUs architecture is an effective method for DNN-based acoustic model learning speed improvement.

27.4 Conclusions

IPA has attracted human–computer interaction field, because speech interface is being used to simplify the interaction between IT devices and users. Most current speech interfaces used DNN-based acoustic model, and its accuracy showed advantages over accuracy of speech interface using GMM-based acoustic model. However, DNN-based acoustic model learning is required to spend much time. For improving learning speed of DNN, previous works used multi-core CPU systems, but there was limitation on the number of cores.

For this reason, many researchers got interested in using GPU to use many cores. However, for using GPU effectively, the training data or the size of the parameters in DNN learning had to be reduced for solving CPU-to-GPU bottleneck problem. Therefore, distributed computing methods such as MapReduce or GE have begun to receive attention.

This paper suggested a learning speed improvement using multi-GPUs on DNN-based acoustic model training in Korean IPA. DNN learning involves iterative, stochastic parameter updates, which depend on the previous updates. The proposed method provides multi-GPUs architecture using GE to satisfy features of DNN learning. DNN-based acoustic models are trained by using 5 slaves with 5 GPUs. It was shown that the learning speed becomes 5 times faster on this implementation while maintaining speech recognition rate.

Acknowledgements This research was supported by Basic Science Research Program through the National Research Foundation of Korea (NRF) funded by the Ministry of Science, ICT and Future Planning (No. NRF-2014R1A1A1002197).

References

Allcock B, Bester J, Bresnahan J, Chervenak AL, Foster I, Kesselman C, Meder S, Nefedova V, Quesnel D, Tuecke S (2002) Data management and transfer in high performance computational grid environments. Parallel Comput J 28:749–771

Bourlard H, Morgan N (1994) Connectionist speech recognition: a hybrid approach. Kluwer Academic Publishers, pp 1–263. doi:10.1007/978-1-4615-3210-1

Castellano M, Mastronardi G, Tarricone G (2009) Intrusion detection using neural networks: a grid computing based data mining approach. Neural Inf Process 5864:777–785

Deng L, Acero A, Wang Y, Wang K, Hon H, Droppo J, Mahajan M, Huang XD (2002) A speech-centric perspectives for human–computer interface. In: IEEE Work. Multimed. Signal Process, pp 263–267

Dragan RV (2002) Sun one grid engine enterprise edition software, 1 October, pp 1–3

Forster I, Kesselan C (1999) The grid: blueprint for a new computing infrastructure. Morgan Kaufman, San Francisco

Li X, Wang YY, Tur G (2011) Multi-task learning for spoken language understanding with shared slots. In: Proc. Annu. Conf. Int. Speech Commun. Assoc. (INTERSPEECH 2011), pp 701–704

Oh K-S, Jung K (2004) GPU implementation of neural networks. Pattern Recognit 37:1311–1314. doi:10.1016/j.patcog.2004.01.013

Raina R, Madhavan A, Ng AY (2009) Large-scale deep unsupervised learning using graphics processors. In: Proc. 26th Int. Conf. Mach. Learn, pp 873–880

Reed DA, Mendes CL, Lu C, Foster I, Kesselmann C (2003) The grid 2: blueprint for a new computing infrastructure. Morgan Kaufman, San Francisco

Seltzer ML, Droppo J (2013) Multi-task learning in deep neural networks for improved phoneme recognition. In: Proc. 2013 IEEE Int. Conf. Acoust. Speech Signal Process. (ICASSP), pp 6965–6969

Steinkraus D, Buck I, Simard PY (2005) Using GPUs for machine learning algorithms. In: Proc. Eighth Int. Conf. Doc. Anal. Recognit, pp 1115–1120

Torresen O, Landsverk J (1998) A review of parallel implementations of backpropagation neural networks. In: Parallel Archit. Artif. Neural Networks, pp 25–64

Tur G (2006) Multitask learning for spoken language understanding. In: Proc. 2006 IEEE Int. Conf. Acoust. Speech Signal Process. (ICASSP), pp 585–588

Wang G (2014) Context-dependent acoustic modelling for speech recognition. PhD Thesis, National University of Singapore

Index

G. Geunbae Lee et al. (eds.), *Natural Language Dialog Systems and Intelligent Assistants*, DOI 10.1007/978-3-319-19291-8

Kolibri 360 GmbH
Ferdinand-Jühlke-Straße 7
99095 Erfurt, Deutschland
produktsicherheit@kolibri360.de